Empire State
Mathematics
May to May

Continental Press

Acknowledgments

Illustrator: Cover and title page: Laurie Conley

ISBN 978-0-8454-5562-3

Copyright © 2008 The Continental Press, Inc.

No part of this publication may be reproduced in any form or by any means, electronic, mechanical, photocopying, recording, or otherwise, without the prior written permission of the publisher. All rights reserved. Printed in the United States of America.

Table of Contents

Introduction 5

UNIT 1 Number Systems and Theory, Part 1 7

6.N.1	Lesson 1	Whole Numbers to Trillions 8
6.N.2, 3, 4, 5	Lesson 2	Properties of Addition and Multiplication 12
6.N.6, 8	Lesson 3	Ratios and Rates 16
6.N.7, 9, 10	Lesson 4	Proportions 20
		Number Systems and Theory, Part 1 Review 24

UNIT 2 Number Systems and Theory, Part 2 27

6.N.11	Lesson 1	Percents 28
6.N.20	Lesson 2	Terminating and Repeating Decimals 32
6.N.21	Lesson 3	Equivalent Forms of Rational Numbers 36
6.N.13, 14, 15	Lesson 4	Ordering Rational Numbers 40
6.N.12	Lesson 5	Applications of Percents 44
		Number Systems and Theory, Part 2 Review 48

UNIT 3 Operations 51

6.N.16	Lesson 1	Adding and Subtracting Fractions 52
6.N.17, 19	Lesson 2	Multiplying and Dividing Fractions 56
6.N.18	Lesson 3	Operations with Mixed Numbers 60
6.N.23, 24, 25	Lesson 4	Exponential Forms 64
6.N.22, 25	Lesson 5	Order of Operations 68
6.N.26, 27	Lesson 6	Estimation with Percents 72
		Operations Review 76

UNIT 4 Algebra 79

5.A.2; 6.A.1	Lesson 1	Variables and Expressions 80
5.A.3; 6.A.2	Lesson 2	Evaluating Expressions 84
5.A.4, 5	Lesson 3	Solving Equations 88
6.A.6	Lesson 4	Evaluating Formulas 92
		Algebra Review 96

UNIT 5 Geometry 99

6.G.1	Lesson 1	Similar Triangles 100
6.G.2, 3	Lesson 2	Area 104
6.G.4	Lesson 3	Volume 108
6.G.5, 6, 7, 9	Lesson 4	Circles 112
6.G.7	Lesson 5	Area of Circles 116

6.G.5, 8	Lesson 6 Central Angles and Sectors	120
5.G.12, 13, 14	Lesson 7 Coordinate Geometry	124
	Geometry Review	128

UNIT 6 Measurement ... 131

6.M.2, 3, 6	Lesson 1 Customary Units of Capacity	132
6.M.4, 5, 6	Lesson 2 Metric Units of Capacity	136
6.M.1	Lesson 3 Relationship of Volume and Capacity	140
6.M.7, 8, 9	Lesson 4 Estimating Measurements	144
	Measurement Review	148

UNIT 6 Statistics ... 151

6.S.5, 6	Lesson 1 Statistics	152
6.S.7, 8	Lesson 2 Bar Graphs	156
6.S.7, 8	Lesson 3 Histograms	160
6.S.7, 8	Lesson 4 Line Graphs	164
6.S.7, 8	Lesson 5 Circle Graphs	168
	Statistics Review	172

UNIT 8 Probability ... 175

5.S.5, 6	Lesson 1 Probability	176
5.S.7	Lesson 2 Sample Spaces	180
6.S.8	Lesson 3 Predictions from Data	184
	Probability Review	188

Empire State Mathematics Grade 6 Supplemental Lessons

Supplemental Lessons for Unit 4 Algebra

6.A.3	Lesson 1 Writing Two-Step Equations	4
6.A.4	Lesson 2 Solving Two-Step Equations	8
6.A.5	Lesson 3 Solving Proportions	12

Supplemental Lesson for Unit 5 Geometry

6.G.10, 11	Lesson 4 More Coordinate Geometry	16

Supplemental Lessons for Unit 8 Probability

6.S.9, 11	Lesson 5 Compound Events	20
6.S.10	Lesson 6 Dependent Events	24

Introduction

This book is called **Empire State Mathematics May to May.** It will help you prepare to take the New York State math test. The math test is given in May. It tests what you have studied from May of last year to May of this year. Like the tests, this book focuses on those skills.

This book reviews what you have already learned in math class. Each four-page lesson starts with a review. Then it gives you problems to practice those skills. The problems are similar to the test questions and even look like them.

At the beginning of each lesson is a list of words you should know. These vocabulary words are placed in a box at the top of the lesson page and marked with this symbol 📖. Be sure you understand these words.

The lesson continues with an explanation of the skills or ideas you need to understand. It reviews a math skill or concept that you have learned in the last year. Examples are worked out for you in numbered steps. At the side are boxes with extra information, marked with this symbol 💼. You can think of these as tools to use. Sometimes these notes explain a way to do something. Other times they tell how one idea relates to another. Or they may remind you of the meaning of a concept.

The next three pages give you practice problems. **Multiple-choice** questions are on the first page. These are problems that give you four answer choices to choose from. The first problem is a Guided Practice question. At the side of the problem is an explanation box that tells how to work out the answer. This box is marked with this symbol ⚙. After the example, there are Independent Practice questions to solve on your own.

Constructed response problems are on the second and third practice pages. These problems ask you to write the answers in your own words. They can take many forms. You may need to show your work or draw a geometric figure. You may be asked to complete a table or graph. Often you will explain how you did something or why an answer is correct.

Some constructed response items are **short-response** questions. These problems have two parts. The first part usually asks you to find the answer to a problem. The second part may ask you to find the answer to another question. Or you may need to explain how you found your answer or why it is correct.

Other constructed response items are **extended-response** questions. These questions are like the short-response questions but have an extra step. Usually they have three parts. They take a little longer to do.

The first page of constructed response problems starts with a Guided Practice question. At the side is a box that explains how to work out the problem. The box is marked with a ⚙. Following this example are Independent Practice questions to solve on your own. The last page presents a problem that is a little longer and more difficult. A boxed 💼 hint tells you how to get started or how to work out some part of this problem.

At the end of each unit is a three-page review section. It mixes problems from all the lessons in the unit. It includes both multiple-choice and constructed response questions. The review questions have no hints or explanations. So they give you the chance to really show what you have learned.

The book concludes with a set of supplemental lessons. These lessons address indicators that were previously taught at grade level but tested the following year. These indicators are now included on the grade 6 test.

Number Systems and Theory, Part 1

Numbers have many different forms and uses. Some numbers are used to describe amounts or measurements. You do that when you name the population of a city or calculate the distance across the United States. Numbers are also used to show ratios and rates, ways of comparing numbers. When you buy packs of 8 hot dogs and bags of 10 hot dog buns, the ratio of hot dogs to buns is 4 to 5. Proportions are equations that show two ratios are equal. In addition to their forms and uses, numbers have certain rules, or properties, that apply to them. For example, the inverse property helps you understand how opposite operations relate to each other.

This unit will help you answer test questions about numbers and how they relate to one another. There are four lessons in this unit:

1. Whole Numbers to Trillions   This lesson reviews place value of whole numbers through trillions. You will also write numbers in standard, word, and expanded forms.

2. Properties of Addition and Multiplication   In this lesson, you will review properties, or rules, that apply when you are adding and multiplying.

3. Ratios and Rates   In this lesson, you will review how to find and use ratios and rates. You will also find and use unit rates.

4. Proportions   This lesson reviews proportions. You will decide if a given proportion is a true statement. You will also solve proportions to find a missing term.

Whole Numbers to Trillions

Indicator 6.N.1

place value   trillions   billions   millions   thousands   ones
periods   standard form   word form   expanded form

The **place value** of each digit in a number depends on its position in that number. **Trillions, billions, millions, thousands,** and **ones** are all place value **periods.**

TRILLIONS	HUNDRED BILLIONS	TEN BILLIONS	BILLIONS	HUNDRED MILLIONS	TEN MILLIONS	MILLIONS	HUNDRED THOUSANDS	TEN THOUSANDS	THOUSANDS	HUNDREDS	TENS	ONES
4,	3	1	8,	6	0	9,	7	2	5,	7	9	4

> Place value periods are separated by commas.
> The periods include:
> trillions
> billions
> millions
> thousands
> ones

The number in this place value chart is read as 4 trillion, 318 billion, 609 million, 725 thousand, 794.

Numbers in standard, or numeric, form can be written in word form. Numbers in written form can be named in standard form.

How is the number 5,080,000,401,000 written in word form?

1. Identify the non-zero digits in the number: 5, 8, 4, and 1.

2. Find the place values of each non-zero digit and write them using words: five trillion, eighty billion, four hundred one thousand.

What is the standard form of the number seventy-nine billion, three hundred seventeen million, forty-five?

1. List the numbers as digits in order according to place value periods separated by commas: billions (79), millions (317), thousands (no digits), ones (45).

2. Write the digits in standard form: 79,317,000,045.

> There are several ways to name numbers.
>
> • **standard form**
> 5,700,230
>
> • **word form**
> five million, seven hundred thousand, two hundred thirty
>
> • **expanded form**
> 5,000,000 + 700,000 + 200 + 30
> or
> (5 × 1,000,000) +
> (7 × 100,000) +
> (2 × 100) + (3 × 10)

UNIT 1 Number Systems and Theory, Part 1

GUIDED PRACTICE

Try this sample multiple-choice problem.

S What is the expanded form of thirty-six billion, nine hundred thousand?

A 3 × 10,000,000 + 6 × 1,000,000 + 9 × 10,000

(B) 3 × 100,000,000 + 6 × 10,000,000 + 9 × 1,000

C 3 × 1,000,000,000 + 6 × 100,000,000 + 9 × 1,000,000

D 3 × 10,000,000,000 + 6 × 1,000,000,000 + 9 × 100,000

> This problem asks you to identify the expanded form of a number written in word form. Identify the digits in each place value period: 36 billions, 900 thousands. Next, write the digits as products of their place values: 36 billions = 3 × 10,000,000,000 + 6 × 1,000,000,000; 900 thousands = 9 × 100,000. The correct answer is D.

INDEPENDENT PRACTICE

Read each problem. Circle the letter of the best answer.

1 Which number is the same as 7 × 100,000,000 + 2 × 1,000,000 + 5 × 100,000?

A 700,200,500
B 700,250,000
(C) 702,500,000
D 725,000,000

2 What is the standard form for three trillion, eight billion, sixty-one million, thirteen?

A 3,008,061,013
B 3,800,610,013
(C) 3,008,061,000,013
D 3,800,610,000,013

3 What is 1,230,000,000 in word form?

A one trillion, twenty-three billion
B one billion, twenty-three million
C one trillion, two hundred thirty billion
D one billion, two hundred thirty million

4 An Internet search produced 478,000,000 results. How is this number written in word form?

A four hundred seventy-eight billion
B four hundred seventy-eight million
C four hundred seventy-eight trillion
D four hundred seventy-eight thousand

5 What is 1,132,000,000 in expanded form?

A 1 × 1,000,000,000 + 1 × 100,000,000 + 3 × 10,000,000 + 2 × 1,000,000

B 1 × 10,000,000 + 1 × 1,000,000 + 3 × 100,000 + 2 × 1,000

C 1 × 100,000,000 + 1 × 1,000,000 + 3 × 100,000 + 2 × 10,000

D 1 × 1,000,000,000 + 1 × 100,000,000 + 3 × 10,000 + 2 × 1,000

UNIT 1 Number Systems and Theory, Part 1

GUIDED PRACTICE

Try this sample constructed response problem.

S One year, the number of tourists visiting New York City was about 43,800,000.

Part A: How is this number written in word form?

Answer: <u>forty-three million, eight hundred thousand</u>

Part B: That same year, the visitors to the city spent about twenty-four billion, seven hundred ten million dollars. How is this number written in standard form?

Answer: <u>$24,710,000,000</u>

> Part A asks you to write 43,800,000 in word form. Think of the names of the periods that are separated by commas, starting from the right. There are 0 ones, 800 thousands, and 43 millions. The correct answer is forty-three million, eight hundred thousand. Part B asks you to write the standard form for a number in word form. Identify the digits in each place value period: 24 billion is the same as 24,000,000,000 and 710 million is the same as 710,000,000. Add these together. The correct answer is 24,710,000,000.

INDEPENDENT PRACTICE

Read the problem. Write your answers.

6 Colby wrote the number 5,000,620,004,000 on the chalkboard.

Part A: Write the expanded form of this number.

Answer: _____

Part B: Write the word form of this number.

Answer: _____

UNIT 1 Number Systems and Theory, Part 1

INDEPENDENT PRACTICE

Read the problem. Write your answers.

7 Each month, the United States government prints about 1,140,000,000 paper bills. The value of this money is more than twenty-two billion dollars.

Part A: How is 1,140,000,000 written in expanded form and in word form?

> Think about the place value of each digit in the number.

Expanded Form: _____

_____

Word Form: _____

_____

Part B: How is twenty-two billion written in standard form?

Answer: _____

UNIT 1 Number Systems and Theory, Part 1

Lesson 2: Properties of Addition and Multiplication

Indicators 6.N.2, 3, 4, 5

commutative property   distributive property   addend   identity property of addition
identity property of multiplication   zero product property   inverse property
opposite operations

The **commutative property** says that numbers can be added or multiplied in any order and the result will be the same.

$$5 + 9 = 14 \qquad 3 \times 7 = 21$$
$$9 + 5 = 14 \qquad 7 \times 3 = 21$$

The commutative property is *not* true for subtraction or division.

$$6 - 1 \neq 1 - 6$$
$$10 \div 2 \neq 2 \div 10$$

The **distributive property** of multiplication over addition says that the product of a number and a sum is the same as the sum of the products of the number and each addend.

$$8 \times (3 + 2) = (8 \times 3) + (8 \times 2)$$
$$8 \times 5 = 24 + 16$$
$$40 = 40$$

An **addend** is a number that is being added.

The **identity property of addition** says that any number can be added to 0 and the result will be that number. The **identity property of multiplication** says that any number can be multiplied by 1 and the result will be that number.

$$46 + 0 = 46 \qquad 3\tfrac{1}{4} \times 1 = 3\tfrac{1}{4}$$

The **zero product property** says that the product of any number and 0 is always 0.

$$6.173 \times 0 = 0$$
$$0 \times \tfrac{7}{24} = 0$$

The **inverse property** says that **opposite operations** can be used to "undo" each other. Addition and subtraction are opposite operations. Multiplication and division are opposite operations.

$$8 - 3 = 5 \text{ because } 5 + 3 = 8$$
$$36 \div 9 = 4 \text{ because } 4 \times 9 = 36$$

UNIT 1 Number Systems and Theory, Part 1

GUIDED PRACTICE

Try this sample multiple-choice problem.

S Which expression has the same value as 7 × 312?

- A (7 × 300) + 12
- B (7 × 300) × 12
- C (7 × 300) + (7 × 12)
- D (7 × 300) × (7 × 12)

> This problem asks you to identify an equivalent expression for 7 × 312. Any number can be expressed as the sum of other numbers: 312 = 300 + 12. So, 7 × 312 = 7 × (300 + 12). Apply the distributive property, which says that addends can be multiplied as a sum or separately and the result is the same: 7 × 312 = (7 × 300) + (7 × 12). The correct answer is C.

INDEPENDENT PRACTICE

Read each problem. Circle the letter of the best answer.

1 Which expression has the same product as 2.4 × 9?

- A 2.4 + 9
- B 2.9 × 4
- C 9 × 2.4
- D 9 × 2 + 0.4

2 Which property is shown by this number sentence?

$$35 + 0 = 0 + 35$$

- A zero property
- B distributive property
- C identity property of addition
- D commutative property of addition

3 Which number sentence is true?

- A 3 + 0 = 0
- B 0 × 3 = 0
- C 1 × 3 = 1
- D 3 × 0 = 3

4 Which number sentence correctly shows the distributive property of multiplication over addition?

- A (3 + 6) × 2 = (3 × 2) + (6 × 2)
- B (3 + 6) × 2 = (3 + 2) × (6 + 2)
- C (3 + 6) × 2 = (3 × 6) + (3 × 2)
- D (3 + 6) × 2 = (3 + 6) × (3 + 2)

5 Glen added a number to 5. Then he divided the result by 2. Which sequence of operations could he use to get back to his original number?

- A add 5, then divide by 2
- B divide by 2, then add 5
- C subtract 5, then multiply by 2
- D multiply by 2, then subtract 5

UNIT 1 Number Systems and Theory, Part 1

GUIDED PRACTICE

Try this sample constructed response problem.

S Abigail wrote this number sentence to apply the distributive property.

$$15 \bigcirc 125 = 15 \times 100 + 15 \times \square$$

Part A: What number belongs inside the square?

Answer: ____25____

Part B: What operation belongs inside the circle?

Answer: ____multiplication____

> Part A asks you to identify the missing number that goes in the square. The distributive property says that the product of a number and a sum is the same as the sum of the products of the number and each addend. The number 125 is written as the sum of 100 and the missing number. Since $125 - 100 = 25$, the missing number is 25. The correct answer is 25. Part B asks you to identify the missing operation in the number sentence. The left side shows the product of a number 15 and a sum: 125 can be written as the sum of 100 and 25. So the product symbol is missing. The correct answer is multiplication.

INDEPENDENT PRACTICE

Read the problem. Write your answers.

6 Marni writes two number sentences. In the first, she adds 0 to the number $8\frac{1}{2}$. In the second, she multiplies $8\frac{1}{2}$ by 0.

Part A: What is the result of each number sentence?

Adding 0: _____

Multiplying by 0: _____

Part B: Which properties did you use to find your answers?

Adding 0: _____

Multiplying by 0: _____

UNIT 1 Number Systems and Theory, Part 1

INDEPENDENT PRACTICE

Read the problem. Write your answers.

7 Tyra bought 2 toothbrushes and 2 tubes of toothpaste. Each toothbrush cost $3 and each tube of toothpaste cost $4.

Part A: Show two different ways that Tyra can find the total cost of the toothbrushes and the tubes of toothpaste.

Way 1: _____

Way 2: _____

Part B: What property is used to show that both ways give the same result?

Answer: _____

> Think about the number properties that involve both addition and multiplication.

What is the total cost?

Answer: _____

UNIT 1 Number Systems and Theory, Part 1

Ratios and Rates

Indicators 6.N.6, 8

ratio   rate   unit rate

A **ratio** is a comparison of two quantities.

A box contains 8 plastic spoons and 12 plastic forks. What is the ratio of spoons to forks?

1. Identify the number of spoons and the number of forks. There are 8 spoons and 12 forks.

2. Write the ratio in one of three ways: 8 to 12, 8:12, or $\frac{8}{12}$.

3. Reduce a ratio to lowest terms to make it easier to compare. In lowest terms, the ratio of spoons to forks is $\frac{2}{3}$.

The order of quantities in a ratio is important.

The ratio 2:7 is **not** the same as the ratio 7:2.

Ratios can be written in one of three ways:
- with the word *to*
  5 to 3
- with a colon (:)
  5:3
- as a fraction
  $\frac{5}{3}$

A **rate** is a type of ratio that compares two quantities of different units of measure. A **unit rate** is a comparison of a quantity to a unit of 1.

Some rates are also written as fractions with the numerator and denominator labeled.

Louise earned $45 for working 5 hours. Write the amount she earned as a unit rate.

1. Identify the rate: $45 for 5 hours, or $\frac{\$45}{5 \text{ hours}}$.

2. Rewrite the rate as a comparison of dollars to 1 hour. Divide each unit by 5: $\frac{\$45 \div 5}{5 \text{ hours} \div 5} = \frac{\$9}{1 \text{ hour}}$. The unit rate is $9 an hour.

To find equivalent ratios, divide each quantity by the same number.

Some words or phrases used to indicate rates include: *a, an, every, for each, in, per, to*

The word *per* means "for each."

3 lb *per* box
means
3 lb *for each* box

UNIT 1 Number Systems and Theory, Part 1

GUIDED PRACTICE

Try this sample multiple-choice problem.

S Devon hiked 4 miles in 2 hours. Which describes the unit rate, or speed, in which Devon hiked?

A $\frac{1 \text{ hour}}{2 \text{ miles}}$ C $\frac{2 \text{ hours}}{4 \text{ miles}}$

B $\frac{2 \text{ miles}}{1 \text{ hour}}$ D $\frac{4 \text{ miles}}{2 \text{ hours}}$

> This problem asks you to identify the unit rate. A rate compares two quantities of different units. A unit rate is a type of rate in which the second quantity compared has a rate of 1 unit. As a fraction, a unit rate has a denominator of 1. The only answer choice that is a unit rate is $\frac{2 \text{ miles}}{1 \text{ hour}}$. The correct answer is B.

 INDEPENDENT PRACTICE

Read each problem. Circle the letter of the best answer.

1 Joyce filled a bathtub with 40 gallons of water in 8 minutes. Which describes the rate at which the water filled the bathtub?

A 4 gallons in 80 minutes

B 8 gallons in 40 minutes

C 40 gallons in 8 minutes

D 80 gallons in 4 minutes

2 Which situation describes a rate?

A In a box of 12 eggs, 3 are broken.

B It rained on 7 of the past 10 days.

C There are 6 full baskets and 4 empty baskets.

D There are 2 teachers for every 15 students.

3 Which fraction below represents a ratio that is **not** a rate?

A $\frac{1 \text{ week}}{\$100}$ C $\frac{18 \text{ miles}}{1 \text{ gallon}}$

B $\frac{15 \text{ red bicycles}}{20 \text{ blue circles}}$ D $\frac{50 \text{ newspapers}}{30 \text{ minutes}}$

4 Which situation describes a unit rate?

A Beth spent $6 for each case of soda.

B A train traveled 110 miles in 2 hours.

C Ari got 18 out of 20 problems correct.

D In a box of 16 pencils, 12 are sharpened and the rest are not.

5 Which of these does **not** describe a rate?

A $0.59 a pound

B 7 out of 10 students

C 2 beds in every room

D 10 vacation days each year

UNIT 1 Number Systems and Theory, Part 1

GUIDED PRACTICE

Try this sample constructed response problem.

S A 4-lb bag of apples costs $5. A 3-lb bag of pears costs $4.

Part A: What is the unit rate as a cost per pound for the bag of apples?

Answer: _____$1.25 per pound_____

Part B: Which bag of fruit is the better buy—the apples or the pears?

Answer: _____apples_____

> Part A asks you to identify the unit rate, or cost per pound, for the apples. As a rate, the apples cost $5 for every 4 pounds. To find the unit rate, divide each number by 4: $5 ÷ 4 = $1.25 and 4 ÷ 4 = 1. The correct answer is $1.25 per pound. Part B asks you to identify which fruit is the better buy. The fruit with the lower unit rate is the better buy. The unit rate for the pears is $4 ÷ 3 = $1.33 per pound. The apples have a unit rate of $1.25 per pound, which is lower. The correct answer is apples.

INDEPENDENT PRACTICE

Read the problem. Write your answers.

6 A photocopier can make 100 copies in 4 minutes.

What is the unit rate, or speed, in which this photocopier makes copies?

Answer: _____

Explain why your answer is correct.

_____
_____
_____
_____

UNIT 1 Number Systems and Theory, Part 1

INDEPENDENT PRACTICE

Read the problem. Write your answers.

7 A package contains a total of 12 cracker sandwiches. Of these, 9 have cheese and the rest have peanut butter.

Part A: What is the ratio of crackers with cheese to crackers with peanut butter?

Answer: _____

Part B: Is this ratio also a rate?

Answer: _____

A ratio compares two quantities.

Explain why your answer is correct.

_____
_____
_____
_____

LESSON 4: Proportions

Indicators 6.N.7, 9, 10

proportion ratio cross products means extremes cross multiplying

A **proportion** is a statement that shows two ratios are equal.

$$\frac{4}{5} = \frac{12}{15} \qquad 4:5 = 12:15$$

> A **ratio** is a comparison of two quantities.

This proportion says that the ratio 4 to 5 is the same as the ratio 12 to 15.

Is the proportion $\frac{3}{4} = \frac{9}{16}$ true?

1. Find the **cross products.** To do this, multiply the **means** of the proportion: $4 \times 9 = 36$. Then multiply the **extremes** of the proportion: $3 \times 16 = 48$.

2. Compare the products of the means and the extremes. If they are equal, then the proportion is true: $36 \neq 48$. So the proportion is not true.

> A proportion has means and extremes.
>
>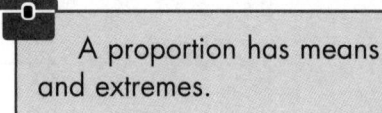
> ↑ ↑
> Means Extremes
>
> Extremes
> ↓ ↓
> 4:5 = 12:15
> ↑ ↑
> Means

You can find the missing term in a proportion by using cross products, or **cross multiplying.**

Amanda uses 6 tomatoes to make 2 servings of tomato sauce. How many tomatoes will Amanda need to make 8 servings of tomato sauce?

1. Set up a proportion. Use the letter t to stand for the number of tomatoes.

   $$\frac{\text{Tomatoes} \rightarrow}{\text{Servings} \rightarrow} \frac{6}{2} = \frac{t}{8} \frac{\leftarrow \text{Tomatoes}}{\leftarrow \text{Servings}}$$

2. Cross multiply: $2 \times t = 6 \times 8$, so $2t = 48$.

3. Divide each side by 2 to find t: $2t \div 2 = 48 \div 2$, so $t = 24$. Amanda needs 24 tomatoes to make 8 servings of tomato sauce.

> Cross multiplying is the same as setting the product of the means equal to the product of the extremes.

> In a proportion, be sure the quantities being compared are in the correct place.
>
> $$\frac{\text{Cost}}{\text{Weight}} = \frac{\$10}{2 \text{ lb}} = \frac{c}{5 \text{ lb}}$$

UNIT 1 Number Systems and Theory, Part 1

GUIDED PRACTICE

Try this sample multiple-choice problem.

S Which set of ratios forms a proportion?

A $\frac{1}{3}$ and $\frac{3}{10}$ C $\frac{2}{5}$ and $\frac{5}{10}$

B $\frac{3}{4}$ and $\frac{6}{9}$ D $\frac{5}{8}$ and $\frac{10}{16}$

> This problem asks you to identify the ratios that form a proportion. Find the products of their means and their extremes. These are equal in a proportion. In choice A, $9 \neq 10$. In choice B, $24 \neq 27$. In choice C, $25 \neq 20$. In choice D, $80 = 80$, so the ratios form a proportion. The correct answer is D.

INDEPENDENT PRACTICE

Read each problem. Circle the letter of the best answer.

1 Which equation can be used to show that $\frac{6}{8} = \frac{30}{40}$?

A $6 \times 8 = 30 \times 40$

B $6 \times 30 = 8 \times 40$

C $6 \times 40 = 8 \times 30$

D $6 + 30 = 40 - 8$

2 Which of the following proportions is true?

A $\frac{5}{3} = \frac{20}{12}$ C $\frac{6}{8} = \frac{16}{18}$

B $\frac{6}{3} = \frac{12}{9}$ D $\frac{8}{5} = \frac{25}{15}$

3 Which proportion is true when $n = 5$?

A $\frac{n}{6} = \frac{12}{20}$ C $\frac{1}{n} = \frac{5}{15}$

B $\frac{n}{4} = \frac{20}{16}$ D $\frac{3}{n} = \frac{15}{20}$

4 What value of k makes this proportion true?

$$\frac{k}{3} = \frac{18}{9}$$

A 6

B 9

C 12

D 54

5 Chuck can read 20 pages in 1 hour. He sets up this proportion.

$$\frac{20}{60} = \frac{50}{x}$$

What does the value of x stand for?

A the time it takes to read 50 pages

B the time it takes to read 60 pages

C the number of pages read in 50 minutes

D the number of pages read in 60 minutes

UNIT 1 Number Systems and Theory, Part 1

GUIDED PRACTICE

Try this sample constructed response problem.

S The ratio $\frac{2}{3}$ compares the number of boys to girls in Mr. Dupont's class.

Part A: Does the ratio $\frac{8}{12}$ also represent the number of boys to girls in this class?

Answer: ___yes___

Part B: If there are 6 boys in the class, how many girls are in the class?

Answer: ___9___

> Part A asks you to determine if $\frac{2}{3}$ and $\frac{8}{12}$ are equivalent. Find the product of the means and the product of the extremes. They should be equal: means—$3 \times 8 = 24$; extremes—$2 \times 12 = 24$. They are equal, so the two ratios are equivalent. Part B asks you to find the number of girls if there are 6 boys. Set up a proportion: $\frac{boys}{girls} = \frac{2}{3} = \frac{6}{g}$. Cross multiply: $3 \times 6 = 2 \times g$. So, $18 = 2g$ and $g = 9$. The correct answer is 9.

INDEPENDENT PRACTICE

Read the problem. Write your answers.

6 A box of 2 water filters costs $8. Each water filter costs the same amount.

Part A: Write a proportion that can be used to find c, the cost of a box of 5 water filters.

Answer: _____

Part B: What is the cost, c, of the 5 water filters?

Answer: _____

22 UNIT 1 Number Systems and Theory, Part 1

INDEPENDENT PRACTICE

Read the problem. Write your answers.

7 Lou earns the same dollar amount each hour he works. So far this summer, he earned $120 working 8 hours. Lou's goal is to earn $3,000 by the end of the summer. He set up the proportion below to show how many hours he will need to work in order to reach his goal.

$$\frac{120}{8} = \frac{3,000}{150}$$

Part A: Is this proportion true?

Answer: _____

Find the product of the means and the product of the extremes.

Explain how you know.

_____
_____
_____
_____

Part B: Next week, Lou plans to work a total of 30 hours. How much money can he expect to earn next week?

Show your work.

Answer: _____

UNIT 1 Number Systems and Theory, Part 1

Number Systems and Theory, Part 1 Review

Read each problem. Circle the letter of the best answer.

1 What property is demonstrated by the number sentence below?

$$9 + 1 = 1 + 9$$

A distributive property

B commutative property

C identity property of addition

D inverse property of addition

2 Which number goes in the box to make this number sentence true?

$$\square \times 3 = 0$$

A -3

B 0

C $\frac{1}{3}$

D 1

3 What is the expanded form of the number 6,003,702,000?

A $6 \times 1,000,000 + 3 \times 100,000 + 7 \times 10,000 + 2 \times 1,000$

B $6 \times 1,000,000 + 3 \times 1,000,000 + 7 \times 100,000 + 2 \times 1,000$

C $6 \times 1,000,000,000 + 3 \times 1,000,000 + 7 \times 100,000 + 2 \times 10,000$

D $6 \times 1,000,000,000 + 3 \times 1,000,000 + 7 \times 100,000 + 2 \times 1,000$

4 Which number sentence shows the identity property of multiplication?

A $8 \times 1 = 8$

B $10 \times 0 = 0$

C $7 \times 3 = 3 \times 7$

D $6 \times 1 = 1$

5 Of 640 students, 2 out of 5 ordered hot lunch yesterday. Which proportion could be used to find h, the number of students who ordered hot lunch yesterday?

A $\frac{2}{5} = \frac{640}{h}$

B $\frac{2}{640} = \frac{5}{h}$

C $\frac{h}{5} = \frac{2}{640}$

D $\frac{h}{640} = \frac{2}{5}$

6 Which expression has the same value as 107×17?

A $7 + (100 \times 17)$

B $(100 \times 7) + 17$

C $(100 \times 17) + (7 \times 17)$

D $(100 \times 7) + (100 \times 17)$

7 Chloe subtracted 3 from her age and then multiplied the result by 5. Which sequence of operations could be used to return to Chloe's age?

A add 3, then divide by 5

B divide by 5, then add 3

C subtract 3, then multiply by 5

D multiply by 5, then subtract 3

8 Which proportion is true when $p = 8$?

A $\frac{6}{p} = \frac{9}{12}$

B $\frac{p}{16} = \frac{16}{8}$

C $\frac{15}{9} = \frac{12}{p}$

D $\frac{18}{12} = \frac{p}{4}$

Number Systems and Theory, Part 1 Review

Read each problem. Write your answers.

9 The United States national debt increases an average of 1.5 billion dollars each day.

Part A: How is 1.5 billion written in standard notation?

Answer: _____

Part B: The national debt is over $9,350,000,000,000. How is this amount written in words?

Answer: _____
_____

10 The table at the right shows the costs and bottle sizes for different flavors of juice.

JUICE PRICES

Flavor	Cost	Size
Cranberry	$2.00	24 oz
Grape	$2.80	48 oz
Orange	$3.50	56 oz

Part A: Which flavor of juice has the lowest cost per ounce?

Answer: _____

Part B: Is the cost per ounce a rate? Explain how you know.

_____
_____
_____
_____

Number Systems and Theory, Part 1 Review

Read the problem. Write your answers.

11 Tina completes 15 homework problems in 25 minutes. Eva completes 20 homework problems in 30 minutes.

   Part A: Do Tina and Eva complete their homework problems at the same rate?

   Answer: _____

   Explain why your answer is correct.

   _____
   _____
   _____

   Part B: Marcella completes all her homework problems at the same rate as Tina. It takes Marcella 15 minutes to complete her homework. How many homework problems did Marcella have?

   Show your work.

   Answer: _____

Number Systems and Theory, Part 2

UNIT 2

Percents are a very common use of numbers. Your teacher might use a percent to show how many problems you got correct on your math test. If you got 48 out of 50 problems correct, you got 96% of the test right. You use percents when you figure out how much tip to leave in a restaurant and how much sales tax you will pay on your new computer game. The same number can be written as a fraction, a decimal, and a percent. You can change a number into an equivalent form to make it easier to compare with other numbers. This also helps you to order numbers correctly.

This unit will help you answer test questions about percents, decimals, and equivalent forms. There are five lessons in this unit:

1. **Percents**   This lesson reviews how to write and find percents.

2. **Terminating and Repeating Decimals**   In this lesson, you will review how to change fractions to decimals. You will also decide if a decimal is terminating or repeating.

3. **Equivalent Forms of Rational Numbers**   This lesson reviews how to write the same number as a fraction, a decimal, and a percent.

4. **Ordering Rational Numbers**   This lesson reviews rational numbers. You will put rational numbers in order, with and without the use of a number line. You will also find the absolute value of numbers.

5. **Applications of Percents**   In this lesson, you will review how to solve different kinds of percent problems, as they relate to real-life situations.

Lesson 1: Percents

Indicator 6.N.11

percent   percent sign (%)   equivalent fractions

A **percent** names part of one hundred.

30 parts of 100 is 30%.
8 parts of 100 is 8%.

What percent of the grid below is shaded?

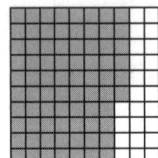

> A percent is written using a **percent sign (%)**.
> 16%   85%

1. Count the total number of small squares in the grid: 100.

2. Count the number of small squares that are shaded: 76. Since 76 squares of 100 are shaded, 76% of the grid is shaded.

You can find the percent of something that is not out of 100 by using proportions or writing equivalent fractions.

Dan made 12 out of 20 free throws during the basketball season. What percent of free throws did Dan make?

1. Write the fraction of free throws made: $\frac{12}{20}$.

2. Set up a proportion using 100 in the denominator: $\frac{12}{20} = \frac{p}{100}$.

3. Solve the proportion for p, the percent. Add a percent sign to the result: $20p = 1{,}200 \rightarrow p = 60$. Dan made 60% of his free throws.

> **Equivalent fractions** have the same value. They can be simplified to the same fraction.
> $\frac{2}{3}$ and $\frac{8}{12}$ are equivalent because $\frac{8 \div 4}{12 \div 4} = \frac{2}{3}$.

> Two fractions form a proportion if their cross products are equal.

UNIT 2 Number Systems and Theory, Part 2

GUIDED PRACTICE

Try this sample multiple-choice problem.

S Lisa planned to ride her bike 8 miles today. She only rode her bike 6 miles. What percent of the miles she planned to ride her bike did Lisa actually ride?

- A 60%
- B 68%
- C 75%
- D 80%

> This problem asks you to find the percent that is equal to 6 out of 8. Set up a proportion using 100 as the denominator of one ratio: $\frac{6}{8} = \frac{p}{100}$. Solve the proportion for p: $8p = 600$, so $p = 75$. Add a percent sign to the result. The correct answer is C.

INDEPENDENT PRACTICE

Read each problem. Circle the letter of the best answer.

1 Tyrone printed sixty-five percent of a document before his printer ran out of paper. What is this percentage written in standard form?

- A 65%
- B 0.65%
- C 6.5%
- D 0.0065%

2 Which fraction has the same value as 9%?

- A $\frac{9}{1}$
- B $\frac{9}{10}$
- C $\frac{9}{100}$
- D $\frac{90}{100}$

3 Naomi slept for 37.5% of a day. How is this percentage written in word form?

- A three hundred seventy-five percent
- B thirty-seven and five-tenths percent
- C thirty-seven and five-hundredths percent
- D three hundred seven and five-tenths percent

4 A recipe calls for $\frac{1}{4}$ cup of milk. Which fraction is equivalent to $\frac{1}{4}$?

- A $\frac{14}{100}$
- B $\frac{25}{100}$
- C $\frac{40}{100}$
- D $\frac{44}{100}$

5 Homeroom A has $\frac{7}{20}$ boys. Homeroom B has $\frac{15}{25}$ boys, C has $\frac{17}{25}$ boys, and D has $\frac{14}{20}$ boys. Which homeroom has 70% boys?

- A homeroom A
- B homeroom B
- C homeroom C
- D homeroom D

6 Gisella bought 20 red cups and 30 blue cups for a party. What percent of the total cups bought are blue?

- A 30%
- B 40%
- C 50%
- D 60%

UNIT 2 Number Systems and Theory, Part 2

GUIDED PRACTICE

Try this sample constructed response problem.

S One of the 16 watches a jeweler sold last week was returned.

Part A: What percent of these watches was returned? Write your answer in standard form.

Answer: ___6.25%___

Part B: Write the percentage from part A in word form.

Answer: ___six and twenty-five___
___hundredths percent___

> Part A asks you to find the percent that is equal to 1 out of 16. Write and solve a proportion: $\frac{1}{16} = \frac{p}{100} \to 16p = 100 \to p = 6.25$. The correct answer is 6.25%. Part B asks you to write this percentage in word form. Write the decimal number using words and include the word *percent* at the end. The correct answer is six and twenty-five hundredths percent.

INDEPENDENT PRACTICE

Read the problem. Write your answers.

7 Carl defined seventy-seven and five-tenths percent of the words correctly in a vocabulary puzzle.

Part A: Write this percent in standard form.

Answer: _____

Part B: Stuart defined 36 out of 50 words correctly in this vocabulary puzzle. What percent of the words did Stuart define correctly?

Answer: _____

UNIT 2 Number Systems and Theory, Part 2

Read the problem. Write your answers.

8 A book has 25 chapters. Drake has read 14 chapters so far.

Part A: What percent of the book has Drake read? Write your answer in standard form.

Show your work.

A percent shows a part of 100.

Answer: _____

Part B: Write the percentage from part A in word form.

Answer: _____

Terminating and Repeating Decimals

fraction   decimal   equivalent   numerator   denominator
terminating decimal   convert   repeating decimal

When a **fraction** and a **decimal** name the same number, they are **equivalent.** To find a decimal that is equivalent to a fraction, divide the **numerator** of the fraction by its **denominator.**

A decimal quotient that ends is called a **terminating decimal.**

What is the decimal equivalent of $\frac{9}{12}$?

1. Divide the numerator by the denominator.

$$\frac{9}{12} = 12\overline{)9.00} = 0.75$$
$$\phantom{\frac{9}{12} = 12\overline{)}}\underline{8\;4}$$
$$\phantom{\frac{9}{12} = 12\overline{)99}}60$$
$$\phantom{\frac{9}{12} = 12\overline{)99}}\underline{60}$$

2. Identify the decimal. Since its last digit is 5 hundredths, 0.75 is a terminating decimal. So $\frac{9}{12}$ is equivalent to 0.75.

> The top number in a fraction is the numerator. The bottom number is the denominator.
>
> $\frac{2}{3}$ ← Numerator
> $\phantom{\frac{2}{3}}$ ← Denominator

> To **convert,** or change, a decimal to a fraction, write the digits of the decimal in the numerator and the place value of the decimal in the denominator.
>
> $0.3 = \frac{3}{10}$ $0.625 = \frac{625}{1,000}$

A decimal that repeats a pattern of digits without end is called a **repeating decimal.**

What is the decimal equivalent of $\frac{8}{9}$?

1. Divide the numerator by the denominator.

$$\frac{8}{9} = 9\overline{)8.00} = 0.\overline{8}$$
$$\phantom{\frac{8}{9} = 9\overline{)}}\underline{7\;2}$$
$$\phantom{\frac{8}{9} = 9\overline{)88}}80$$
$$\phantom{\frac{8}{9} = 9\overline{)88}}\underline{72}$$
$$\phantom{\frac{8}{9} = 9\overline{)88}}80\ldots$$

2. Identify the decimal. The digit 8 repeats, so 0.88… is a repeating decimal. Show this by drawing a bar over the digit that repeats. So $\frac{8}{9}$ is equivalent to $0.\overline{8}$.

> If more than one digit repeats in a decimal, place the bar over all the digits that repeat.
>
> $\frac{1}{22} = 0.0454545\ldots = 0.0\overline{45}$

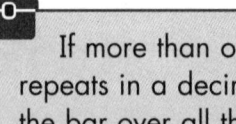

UNIT 2 Number Systems and Theory, Part 2

GUIDED PRACTICE

Try this sample multiple-choice problem.

S Only five-twelfths of the seats on the bus were filled this morning. How is this number written as a decimal?

A 0.416

B 0.41$\overline{6}$

C 0.512

D 0.51$\overline{2}$

> This problem asks you to convert five-twelfths to a decimal. First write this fraction using numbers: $\frac{5}{12}$. Then divide the numerator by the denominator: 5 ÷ 12 = 0.4166.... The digit 6 repeats in this decimal. The correct answer is B.

INDEPENDENT PRACTICE

Read each problem. Circle the letter of the best answer.

1 A clothing store marked down all its sweaters $\frac{1}{5}$ off. How is this fraction written as a decimal?

A 0.15   C 0.25

B 0.2   D 0.4

2 Tim ate three-eighths of a pizza. How is three-eighths written as a decimal?

A 0.3

B 0.333

C 0.375

D 0.38

3 Which of these fractions is also a terminating decimal?

A $\frac{2}{3}$ C $\frac{5}{15}$

B $\frac{4}{9}$ D $\frac{9}{16}$

4 Which decimal has a value between $\frac{1}{4}$ and $\frac{1}{3}$?

A 0.13   C 0.24

B 0.$\overline{18}$ D 0.$\overline{32}$

5 Which of these fractions is also a repeating decimal?

A $\frac{5}{6}$ C $\frac{6}{8}$

B $\frac{5}{8}$ D $\frac{8}{16}$

6 About $\frac{7}{12}$ of the tickets sold to a school play were student tickets. How is this fraction written as a decimal?

A 0.58$\overline{3}$

B 0.$\overline{583}$

C 0.7$\overline{12}$

D 0.712

UNIT 2 Number Systems and Theory, Part 2

GUIDED PRACTICE

Try this sample constructed response problem.

S Tanya wrote two fractions. Each fraction had a denominator of 6. One fraction is equivalent to a repeating decimal. The other fraction is equivalent to a terminating decimal.

   Part A: What could be the numerator of the fraction that is equivalent to the repeating decimal?

   Answer: ____1____

   Part B: What could be the numerator of the fraction that is equivalent to the terminating decimal?

   Answer: ____3____

> Part A asks you to find the numerator of a fraction with a denominator of 6 that is equivalent to a repeating decimal. Start with the number 1. Divide the numerator by the denominator to see if the decimal is repeating: $1 \div 6 = 0.166...$. The 6 repeats, so $\frac{1}{6}$ is a repeating decimal. A correct answer is 1. Part B asks you to find the numerator of a fraction with a denominator of 6 that is equivalent to a terminating decimal, or one that ends. The fraction $\frac{3}{6} = 0.5$, which is a terminating decimal. A correct answer is 3.

INDEPENDENT PRACTICE

Read the problem. Write your answers.

7 Kevin grew $\frac{3}{4}$ inch last year. Marshall grew $\frac{8}{12}$ inch last year.

   Who grew the greater amount last year, Kevin or Marshall?
   Show your work.

   Answer: _____

INDEPENDENT PRACTICE

Read the problem. Write your answers.

8 Ms. Pawley wrote these fractions on the chalkboard.

$$\frac{4}{5} \quad \frac{8}{11} \quad \frac{1}{3} \quad \frac{4}{25} \quad \frac{16}{24}$$

Part A: Which of these fractions are equivalent to repeating decimals?

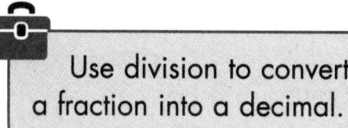
Use division to convert a fraction into a decimal.

Answer: _____

Part B: Write a fraction different from those listed above that is equivalent to a terminating decimal and another that is equivalent to a repeating decimal.

Terminating decimal: _____

Repeating decimal: _____

UNIT 2 Number Systems and Theory, Part 2

Equivalent Forms of Rational Numbers

Indicator 6.N.21

fraction   decimal   percent   equivalent

The same number can be written as a **fraction**, as a **decimal**, and as a **percent**. These are **equivalent** forms of the same number.

$$\frac{4}{10} = \frac{40}{100} = 0.4 = 40\% \qquad \frac{6}{25} = \frac{24}{100} = 0.24 = 24\%$$

> To change a decimal to a fraction, write the digits of the decimal in the numerator and the place value of the decimal in the denominator.
>
> $0.32 = \frac{32}{100} = \frac{8}{25}$

Percents can be changed to decimals by moving the decimal point.

How is 35% written as a decimal?

1. Drop the percent sign: 35% → 35.
2. Move the decimal point two places to the **left:** 35 → 0.35. As a decimal, 35% is 0.35.

> To change a decimal to a percent, move the decimal point two places to the **right** and add a percent sign.

Percents can also be changed to fractions. Recall that a percent is a number out of 100.

How is 35% written as a fraction in lowest terms?

1. Drop the percent sign: 35% → 35.
2. Write the number as a fraction over 100: $\frac{35}{100}$.
3. Simplify: $\frac{35}{100} = \frac{35 \div 5}{100 \div 5} = \frac{7}{20}$. In lowest terms, 35% is $\frac{7}{20}$.

> To change a fraction to a percent, first change the fraction to a decimal. Then move the decimal point to find the percent.
>
> $\frac{1}{5} = 0.2 = 20\%$

UNIT 2 Number Systems and Theory, Part 2

GUIDED PRACTICE

Try this sample multiple-choice problem.

S Amelia and her friends shared $\frac{5}{8}$ of a pack of gum. What percent of the pack of gum did they share?

A 52.5%

B 58%

C 62.5%

D 75%

> This problem asks you to identify the equivalent percent of $\frac{5}{8}$. First change the fraction to a decimal: $\frac{5}{8} = 5 \div 8 = 0.625$. Next, change the decimal to a percent by moving the decimal point two places to the right and adding a percent sign: $0.625 \rightarrow 62.5\%$. The correct answer is C.

INDEPENDENT PRACTICE

Read each problem. Circle the letter of the best answer.

1 Which number has the same value as 0.8?

A 8%

B $\frac{1}{8}$

C $\frac{88}{100}$

D 80%

2 Of the plants in Victor's garden, $\frac{3}{12}$ are tomatoes. How is this fraction written as a decimal?

A 0.25

B 0.3

C 0.312

D 0.4

3 The price of a new patio set is discounted by 33% this weekend. How is this percent written as a decimal?

A 3.3

B 0.33

C 33.0

D 0.0033

4 Which of the following number sentences is true?

A $15\% = \frac{1}{5}$

B $25 = \frac{25}{100}$

C $\frac{3}{5} = 60\%$

D $0.7 = 0.7\%$

5 Pauline walks 0.85 mile to school. What fraction of a mile is this?

A $\frac{3}{4}$

B $\frac{5}{8}$

C $\frac{11}{12}$

D $\frac{17}{20}$

6 Which number is **not** the same as $\frac{1}{20}$?

A 2%

B 5%

C 0.05

D $\frac{2}{40}$

UNIT 2 Number Systems and Theory, Part 2

GUIDED PRACTICE

Try this sample constructed response problem.

S In a survey, 48% of the responders said they go to the local library each month.

Part A: How is 48% written as a decimal?

Answer: __0.48__

Part B: How is 48% written as a fraction in lowest terms?

Answer: $\frac{12}{25}$

> Part A asks you to change 48% to a decimal. Drop the percent sign and move the decimal point two places to the left: 48% → 0.48. The correct answer is 0.48. Part B asks you to change 48% to a fraction. Drop the percent sign and write the number as a fraction with a denominator of 100: $48\% = \frac{48}{100}$. Simplify by dividing the numerator and denominator by the same largest number possible: $\frac{48 \div 4}{100 \div 4} = \frac{12}{25}$. The correct answer is $\frac{12}{25}$.

INDEPENDENT PRACTICE

Read the problem. Write your answers.

7 A bookmark cost 0.9 of a dollar.

Part A: What fraction of a dollar is 0.9? Write your answer in lowest terms.

Answer: _____

Part B: What percent of a dollar is 0.9?

Answer: _____

UNIT 2 Number Systems and Theory, Part 2

INDEPENDENT PRACTICE

Read the problem. Write your answers.

8 Chiara found 16 coins when cleaning her room. Of these coins, 6 were quarters.

Part A: What fraction of the coins that Chiara found were quarters? Write your answer in lowest terms.

Answer: _____

Part B: How is this fraction written as a decimal and as a percent?

Decimal: _____

Percent: _____

> Move the decimal point to the right when changing from a decimal to a percent.

UNIT 2 Number Systems and Theory, Part 2

Ordering Rational Numbers

Indicators 6.N.13, 14, 15

rational number   integers   absolute value   bars

A **rational number** is any number that can be written as a fraction in the form $\frac{p}{q}$, where p and q are **integers.** Rational numbers include fractions, as well as all integers, terminating decimals, and repeating decimals since they can also be written in fraction form. Here are some examples of rational numbers:

> Integers include all whole numbers and their opposites.
> ..., -3, -2, -1, 0, 1, 2, 3, ...

-15 $-\frac{6}{7}$ 3.2   0 $\frac{1}{75}$ 8   -2.5555...

Rational numbers can be placed in order on a number line.

List these rational numbers in order from **least to greatest.**

$\frac{4}{9}$ $-2\frac{2}{5}$ 1.5   -2.25   1

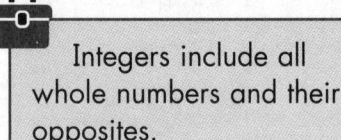

> Rational numbers are easy to compare when they are changed to equivalent forms first.

1. Draw a number line. Place each rational number on it.

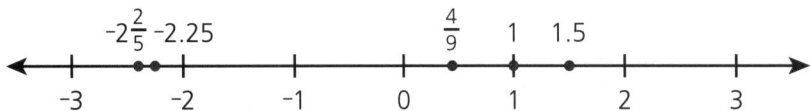

2. Compare the numbers. On a number line, the numbers on the left are smaller than the numbers on the right:
$-2\frac{2}{5} < -2.25 < \frac{4}{9} < 1 < 1.5$.

The **absolute value** of a number is its distance from 0 on a number line.

What is $|-4.5|$, the absolute value of -4.5?

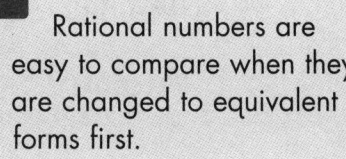

> Absolute values are represented using **bars.**
> $|-2|$ is read as "the absolute value of negative two."

1. Locate -4.5 on a number line.

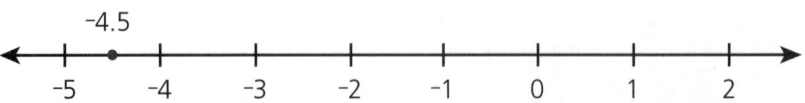

2. Count its distance from 0: 4.5 units. So $|-4.5| = 4.5$.

40   UNIT 2 Number Systems and Theory, Part 2

© The Continental Press, Inc.   Do not duplicate.

GUIDED PRACTICE

Try this sample multiple-choice problem.

S Look at these rational numbers

$$-0.1 \quad \frac{2}{10} \quad -\frac{1}{9} \quad 0.25$$

Which list shows these numbers in order from **least to greatest?**

A $-0.1, -\frac{1}{9}, \frac{2}{10}, 0.25$

B $-\frac{1}{9}, -0.1, 0.25, \frac{2}{10}$

C $-\frac{1}{9}, -0.1, \frac{2}{10}, 0.25$

D $-0.1, \frac{2}{10}, -\frac{1}{9}, 0.25$

> This problem asks you to order numbers from least to greatest. First change each fraction to a decimal to make it easier to compare: $\frac{2}{10} = 0.2$ and $-\frac{1}{9} = -0.1111...$ or $-0.\overline{1}$. Negative numbers are smaller than positive numbers. The smallest number is $-0.\overline{1}$ or $-\frac{1}{9}$. The next smallest number is -0.1. Since 0.2 is smaller than 0.25, the largest number is 0.25. The correct answer is C.

INDEPENDENT PRACTICE

Read each problem. Circle the letter of the best answer.

1 What is the absolute value of 8.1?

A 8.1   C -8.1

B $\frac{1}{8.1}$ D $-\frac{1}{8.1}$

2 Which number is closest to point *P* on this number line?

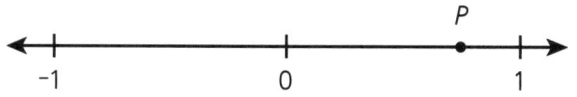

A $\frac{1}{2}$ C $\frac{3}{4}$

B $\frac{2}{5}$ D $\frac{9}{10}$

3 Which number sentence is true?

A $|-6| < |-2|$ C $|-8| < |-5|$

B $|-6| > |2|$ D $|5| > |-8|$

4 Which statement is true?

A Absolute values can be positive or negative numbers.

B Only integers and whole numbers have absolute values.

C The absolute value of any number is always its opposite.

D A number and its opposite have the same absolute value.

5 Which number sentence below is true?

A $-\frac{3}{8} > -\frac{1}{4}$ C $-\frac{4}{5} < -\frac{2}{3}$

B $-6 > -3$ D $-1.125 < -1.4$

UNIT 2 Number Systems and Theory, Part 2

GUIDED PRACTICE

Try this sample constructed response problem.

S Farina took the absolute values of each of these numbers.

$$-3.25 \quad 2\frac{1}{5} \quad -3$$

Part A: List the absolute values of these numbers in order from **least to greatest.**

Answer: $|2\frac{1}{5}|, |-3|, |-3.25|$

Part B: Draw and label the absolute values of these numbers on this number line.

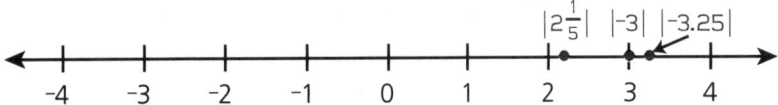

> Part A asks you to order the absolute values of three numbers. First find the absolute values: $|-3.25| = 3.25$, $|2\frac{1}{5}| = 2\frac{1}{5}$, and $|-3| = 3$. In order, the absolute values are $2\frac{1}{5}$, 3, and 3.25. The correct answer is $|2\frac{1}{5}|, |-3|, |-3.25|$. Part B asks you to place the absolute values of these numbers on a number line. All are positive values between 2 and 4. The correct answer is shown at the left.

INDEPENDENT PRACTICE

Read the problem. Write your answers.

6 The lengths of some pencils are shown below.

$$\frac{11}{16} \text{ in.} \quad \frac{2}{3} \text{ in.} \quad \frac{9}{12} \text{ in.} \quad \frac{5}{8} \text{ in.}$$

Part A: Which length is the shortest?

Answer: _____

Part B: Which length is the longest?

Answer: _____

UNIT 2 Number Systems and Theory, Part 2

INDEPENDENT PRACTICE

Read the problem. Write your answers.

7 Look at these numbers.

$$|-1.4| \quad \tfrac{3}{5} \quad -0.2 \quad -\tfrac{1}{1} \quad |1\tfrac{1}{5}|$$

Part A: Draw and label a point for each number on this number line.

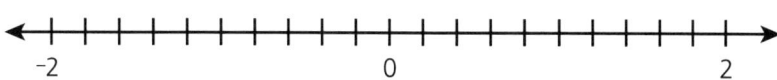

Part B: Which numbers are less than 0?

Answer: _____

Which numbers are more than 1 but less than 2?

Answer: _____

> Write each number in equivalent form to make them easier to compare.

UNIT 2 Number Systems and Theory, Part 2

Applications of Percents

Indicator 6.N.12

percent problems   rate   base

Percent problems have three parts: a percent or part of a number (n), a percentage or **rate** (p), and an original amount or **base** (b).

$$\text{percent of number} = \text{rate} \times \text{base} \qquad n = p\% \times b$$

Multiply to find the percent of a number.

Johanna has 20 ribbons. Of these, 25% are blue. How many blue ribbons does she have?

1. Identify the given parts and write a percent problem: rate = 25%, base = 20, so $n = 25\% \times 20$.

2. Multiply: $25\% \times 20 = 0.25 \times 20 = 5$ blue ribbons.

 The word *of* in a percent problem means multiply. The word *is* means equals.

What percent *of* 24 *is* 6?
$p \times 24 = 6$

 Think: *What number is 25% of 20?*

Divide to find what percent one number is of another.

The school band has 50 members. Of these, 15 are sixth graders. What percent are sixth graders?

1. Identify the given parts and write a percent problem: base = 50, percent of number = 15, so $15 = p \times 50$.

2. Divide and change the decimal to a percent: $15 \div 50 = 0.3 = 30\%$ are sixth graders.

 Think: *15 is what percent of 50?*

 To change a decimal to a percent, move the decimal point two places to the right. Add a percent sign (%).

Divide to find a number when a percent is known.

Damien worked 12 hours last weekend. This is 60% of the total hours he worked that week. How many total hours did Damien work that week?

1. Identify the given parts and write a percent problem: rate = 60%, percent of number = 12, so $12 = 60\% \times b$.

2. Divide: $12 \div 60\% = 12 \div 0.6 = 20$ total hours.

 Think: *12 is 60% of what number?*

GUIDED PRACTICE

Try this sample multiple-choice problem.

S Gordon's lunch bill was $8.00. He also paid a 20% tip. What total amount did Gordon pay for his lunch?

A $8.20

B $8.40

C $9.20

D $9.60

> This problem asks you to find the total amount of a lunch bill, including a 20% tip. First find 20% of $8.00: $n = 20\% \times \$8.00 = 0.2 \times \$8.00 = \$1.60$. The tip amount is $1.60. Add this to the amount of the lunch bill: $8.00 + $1.60 = $9.60. The correct answer is D.

INDEPENDENT PRACTICE

Read each problem. Circle the letter of the best answer.

1 What number is 5% of 40?

A   2   C   5

B   4   D   20

2 Ms. Feldman planted 36 tulip bulbs. This is 60% of all the bulbs she planted. How would you find the total number of bulbs Ms. Feldman planted?

A   divide 36 by 0.6

B   divide 60 by 36

C   multiply 36 by 0.6

D   multiply 60 by 36

3 Sienna traveled a total distance of 400 miles in two days. The first day she traveled 300 miles. What percent of her total distance did Sienna travel the first day?

A   30%   C   70%

B   40%   D   75%

4 Last month, a shoe store sold 550 pairs of footwear. Of these, 28% were boots. How many pairs of boots were sold last month?

A   78   C   154

B   140   D   280

5 Miyuko saved $120 last month. This is 30% of what she earned. What dollar amount did Miyuko earn?

A $36   C $360

B $90   D $400

6 In a survey of 200 people, 48 of them said they watch less than 5 hours of television each week. What percent of people surveyed said they watch less than 5 hours of television each week?

A   10%   C   48%

B   24%   D   96%

UNIT 2 Number Systems and Theory, Part 2

GUIDED PRACTICE

Try this sample constructed response problem.

S Jesse has $350 in a savings account that earns 2% interest each year.

Part A: How much interest will Jesse earn in a year on this amount?

Answer: _____$7_____

Part B: Aiden has a savings account with the same rate of interest. Last year, his money earned $25 in interest. How much money was in Aiden's savings account last year?

Answer: _____$1,250_____

> Part A asks you to find the amount of interest earned in a year on $350. Think: What number is 2% of 350? Solve: $n = 2\% \times 350 = 0.02 \times 350 = 7$. The correct answer is $7. Part B asks you to find the original amount of money in a savings account that earned $25 in interest in one year. Think: 25 is 2% of what number? Solve: $25 = 2\% \times b$; $25 = 0.02 \times b$; $b = 25 \div 0.02 = 1,250$. The correct answer is $1,250.

INDEPENDENT PRACTICE

Read the problem. Write your answers.

7 The Davis family had a yard sale.

- On Friday, they made $150.
- On Saturday, they made $400.
- On Sunday, they made $200.

What percent of the total money did they make on Friday?

Show your work.

Answer: _____

INDEPENDENT PRACTICE

Read the problem. Write your answers.

8 Mr. Chen and Ms. Brown give swimming lessons. Of Mr. Chen's 28 students, 75% are beginning swimmers.

Part A: How many of Mr. Chen's students are beginning swimmers?

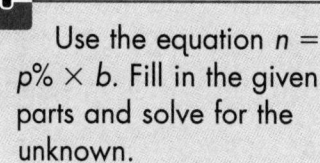

Use the equation $n = p\% \times b$. Fill in the given parts and solve for the unknown.

Answer: _____

Part B: Ms. Brown has a total of 24 students. Of these, 9 are beginning swimmers. What percent of these students are beginning swimmers?

Show your work.

Answer: _____

UNIT 2 Number Systems and Theory, Part 2

Number Systems and Theory, Part 2 Review

Read each problem. Circle the letter of the best answer.

1 Danielle paid four and five-tenths percent sales tax. How is four and five-tenths percent written in standard form?

A 4.5%
B 450%
C 4.05%
D 0.045%

2 Tyler is writing thank-you cards. So far he has written 35% of them. What fraction of cards has Tyler written so far?

A $\frac{3}{5}$
B $\frac{3}{8}$
C $\frac{7}{20}$
D $\frac{9}{20}$

3 Which of these numbers has the smallest value?

A $-|-58|$
B $|-85|$
C $-|76|$
D $|-67|$

4 Of the cars serviced at an auto shop, $\frac{8}{15}$ had oil changes. How is $\frac{8}{15}$ written as a decimal?

A $0.5\overline{3}$
B $0.\overline{53}$
C $0.8\overline{15}$
D $0.\overline{815}$

5 A baker made 150 muffins. Of these, 40% were apple muffins. How many apple muffins did the baker make?

A 40
B 60
C 80
D 100

6 Which number sentence is true?

A $0.15\% = 15$
B $\frac{15}{100} = 15$
C $0.15\% = \frac{0.15}{100}$
D $0.15 = 15\%$

7 Which point on the number line is closest to $-3\frac{1}{3}$?

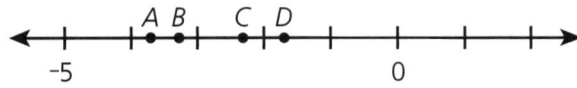

A point A
B point B
C point C
D point D

8 This table shows the areas, in square meters, of different classrooms in a school building.

CLASSROOM AREAS

Classroom	Area (sq m)
A	63.75
B	$63\frac{13}{16}$
C	$63\frac{3}{5}$

Which list shows these classrooms in order from **smallest to largest** area?

A A, B, C
B A, C, B
C C, B, A
D C, A, B

Number Systems and Theory, Part 2 Review

Read each problem. Write your answers.

9. Look at this set of rational numbers.

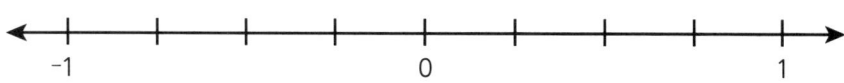

   0.5 $|-\frac{3}{4}|$ $\frac{2}{3}$ $-\frac{2}{5}$ 0.85

   Part A: On the number line below, draw and label a point for each number.

   Part B: Leeza wants to put the number $\frac{7}{12}$ on this number line. Between which two numbers should this number go?

   Answer: _____

10. Kay walked 0.8 mile to the park to meet Nicole.

    Part A: What fraction of a mile did Kay walk? Write your answer in lowest terms.

    Answer: _____

    Part B: Nicole walked $\frac{13}{20}$ mile. How is $\frac{13}{20}$ written as a decimal?

    Answer: _____

Number Systems and Theory, Part 2 Review

Read the problem. Write your answers.

11 A train has 20 passenger cars and 30 freight cars.

Part A: What percent of the total cars are passenger cars?

Answer: _____

Part B: A model train has 12 passenger cars. This is 75% of the total cars. How many total cars does this model train have?

Show your work.

Answer: _____

UNIT 3 Operations

You have already learned how to add, subtract, multiply, and divide whole numbers. These operations are also used when working with fractions and mixed numbers. Some numbers are shown in exponential form, using a base number and an exponent. You follow certain steps to find the value of a number in exponential form. To solve an expression that has more than one operation, you must follow the correct order of operations. Estimation is another skill related to operations. When an exact answer is unnecessary, you can find a good estimate.

This unit will help you answer test questions about operations. There are six lessons in this unit:

1. **Adding and Subtracting Fractions**   This lesson reviews how to add and subtract fractions with unlike denominators. You will find equivalent fractions using the least common denominator.

2. **Multiplying and Dividing Fractions**   This lesson reviews how to multiply and divide fractions with unlike denominators. You will also find the reciprocal of fractions.

3. **Operations with Mixed Numbers**   In this lesson, you will review how to add, subtract, multiply and divide mixed numbers. You will change mixed numbers to improper fractions.

4. **Exponential Forms**   In this lesson, you will review how to write a number in exponential form. You will also find the value of a number written in exponential form.

5. **Order of Operations**   This lesson reviews the correct order of operations to use when finding the value of an expression with more than one operation.

6. **Estimation with Percents**   In this lesson, you will review how to find a good estimate when working with percent problems. You will also decide if an estimate is reasonable for a given situation.

Adding and Subtracting Fractions

LESSON 1

Indicator 6.N.16

unlike denominators   equivalent fractions   like denominators
least common denominator (LCD)   least common multiple (LCM)

To add or subtract fractions with **unlike denominators,** the fractions need to be rewritten as **equivalent fractions** with **like denominators.**

> To add or subtract fractions with like denominators, combine the numerators. The denominator stays the same.
>
> $\frac{2}{7} + \frac{4}{7} = \frac{6}{7}$

Emilio ate $\frac{11}{16}$ pound of salad from a salad bar. Terrence ate $\frac{7}{8}$ pound of salad. How much more salad did Terrence eat than Emilio?

1. Write a subtraction problem: $\frac{7}{8} - \frac{11}{16}$.

2. Rewrite the fractions as equivalent fractions with the same denominator. To do this, find the **least common denominator (LCD)** of the fractions. The LCD is the same as the **least common multiple (LCM)** of the denominators.

   > To find the LCD of two numbers, first list the multiples of each number. Then look for the smallest multiple that is the same for both numbers.

   Multiples of 8: 8, **16,** 24, 32, 40, …
   Multiples of 16: **16,** 32, 48, 64, …

   The smallest multiple common to 8 and 16 is 16. So the LCD is 16: $\frac{7}{8} = \frac{7 \times 2}{8 \times 2} = \frac{14}{16}$. Do not change $\frac{11}{16}$ since 16 is already in the denominator.

3. Subtract the numerators of the fractions with like denominators. The denominator stays the same: $\frac{14}{16} - \frac{11}{16} = \frac{14 - 11}{16} = \frac{3}{16}$. Terrence ate $\frac{3}{16}$ pound more salad than Emilio.

   > Equivalent fractions have the same value but different denominators.

52   UNIT 3 Operations

© The Continental Press, Inc.   Do not duplicate.

GUIDED PRACTICE

Try this sample multiple-choice problem.

S DeShaun drank $\frac{1}{8}$ gallon of orange juice and $\frac{3}{4}$ gallon of water yesterday. How much did DeShaun drink altogether?

- **A** $\frac{5}{8}$ gal
- **B** $\frac{7}{8}$ gal
- **C** $\frac{13}{16}$ gal
- **D** $\frac{15}{16}$ gal

> This problem asks you to find the total amount DeShaun drank. Add $\frac{1}{8}$ and $\frac{3}{4}$. First rewrite the fractions with like denominators by finding the LCD of 4 and 8: 8. So, $\frac{1}{8}$ does not change, and $\frac{3}{4}$ becomes $\frac{6}{8}$. Add: $\frac{1}{8} + \frac{6}{8} = \frac{7}{8}$. The correct answer is B.

INDEPENDENT PRACTICE

Read each problem. Circle the letter of the best answer.

1 Kit grew $\frac{1}{12}$ foot this year and $\frac{1}{6}$ foot last year. How much did she grow both years?

- **A** $\frac{1}{9}$ ft
- **B** $\frac{1}{8}$ ft
- **C** $\frac{1}{4}$ ft
- **D** $\frac{1}{3}$ ft

2 The arrow on a spinner stopped on red $\frac{1}{8}$ of the time. It stopped on green $\frac{1}{2}$ of the time. What fraction of the time did the arrow stop on red or green?

- **A** $\frac{1}{5}$
- **B** $\frac{1}{10}$
- **C** $\frac{3}{4}$
- **D** $\frac{5}{8}$

3 Carter walked $\frac{4}{5}$ mile and jogged $\frac{3}{10}$ mile. How much farther did Carter walk than jog?

- **A** $\frac{1}{2}$ mi
- **B** $\frac{1}{3}$ mi
- **C** $\frac{2}{5}$ mi
- **D** $\frac{7}{10}$ mi

4 On a game board, $\frac{3}{8}$ of the sections are worth 50 points and $\frac{3}{5}$ are worth 10 points. How much more of the game board is worth 10 points than 50 points?

- **A** $\frac{1}{3}$
- **B** $\frac{6}{13}$
- **C** $\frac{6}{40}$
- **D** $\frac{9}{40}$

5 Of the books at a book sale, $\frac{2}{9}$ are adventure and $\frac{1}{6}$ are science fiction. What fraction of the books are either adventure or science fiction?

- **A** $\frac{1}{5}$
- **B** $\frac{7}{18}$
- **C** $\frac{2}{45}$
- **D** $\frac{3}{54}$

UNIT 3 Operations

GUIDED PRACTICE

Try this sample constructed response problem.

S Gerry lifted weights for $\frac{2}{3}$ hour and stretched for $\frac{1}{5}$ hour.

Part A: What fraction of an hour did Gerry lift weights and stretch?

Answer: _____$\frac{13}{15}$_____

Part B: How much more time did Gerry spend lifting weights than stretching?

Answer: _____$\frac{7}{15}$_____ hour

> Part A asks you to find the total fraction of an hour spent lifting weights and stretching. Add $\frac{2}{3}$ and $\frac{1}{5}$. First rewrite the fractions with like denominators by finding the LCD of 3 and 5: 15. So $\frac{2}{3} = \frac{10}{15}$ and $\frac{1}{5} = \frac{3}{15}$. Add the numerators. The denominator stays the same: $\frac{10}{15} + \frac{3}{15} = \frac{13}{15}$. The correct answer is $\frac{13}{15}$. Part B asks you to find the difference in times: $\frac{10}{15} - \frac{3}{15} = \frac{7}{15}$. The correct answer is $\frac{7}{15}$ hour.

INDEPENDENT PRACTICE

Read the problem. Write your answers.

6 Of the students involved with the school play, $\frac{5}{12}$ are acting in it and $\frac{2}{9}$ are designing costumes for it.

What fraction of students are either acting in the school play or designing costumes for it?

Show your work.

Answer: _____

54 UNIT 3 Operations

INDEPENDENT PRACTICE

Read the problem. Write your answers.

7 Katy worked four days last week. This table shows the fraction of total time she worked for the week.

KATY'S WORK SCHEDULE

Day	Fraction of Total Time Worked
Sunday	$\frac{1}{3}$
Wednesday	$\frac{5}{24}$
Friday	$\frac{1}{12}$
Saturday	$\frac{3}{8}$

Part A: What fraction of the total time did Katy work on Sunday and Wednesday?

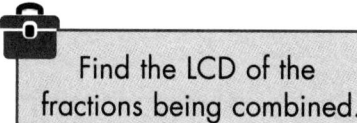
Find the LCD of the fractions being combined.

Answer: _____

Part B: How much more of the total time did Katy work on Saturday than on Friday?

Show your work.

Answer: _____

UNIT 3 Operations

Lesson 2: Multiplying and Dividing Fractions

Indicators 6.N.17, 19

simplest form   common factors   reciprocal   divisor

Fractions with like and unlike denominators are multiplied the same way.

What is $\frac{5}{6} \times \frac{2}{9}$?

1. Multiply the numerators: $5 \times 2 = 10$.
2. Multiply the denominators: $6 \times 9 = 54$.
3. Write the products as a fraction in simplest form: $\frac{10}{54} = \frac{10 \div 2}{54 \div 2} = \frac{5}{27}$.

> To write a fraction in **simplest form,** divide the numerator and denominator by their greatest common factor.
> $\frac{8}{12} = \frac{8 \div 4}{12 \div 4} = \frac{2}{3}$

As a shortcut when multiplying fractions, cancel **common factors.**

What is $\frac{2}{3} \times \frac{7}{8}$?

1. Look for terms with common factors: 2 and 8 have a common factor, 2.
2. Divide terms by the common factor: $\frac{\cancel{2}^{1}}{3} \times \frac{7}{\cancel{8}_{4}}$.
3. Multiply: $\frac{1 \times 7}{3 \times 4} = \frac{7}{12}$.

The **reciprocal** of the **divisor** is used to divide fractions.

What is $\frac{4}{5} \div \frac{2}{3}$?

1. Find the reciprocal of the divisor. The reciprocal of $\frac{2}{3}$ is $\frac{3}{2}$.
2. Multiply by the reciprocal: $\frac{4}{5} \times \frac{3}{2} = \frac{\cancel{4}^{2}}{5} \div \frac{3}{\cancel{2}_{1}} = \frac{6}{5}$.

> To find the reciprocal of a fraction, switch the numerator and denominator.
> The reciprocal of $\frac{3}{7}$ is $\frac{7}{3}$.

UNIT 3 Operations

GUIDED PRACTICE

Try this sample multiple-choice problem.

S Tito has $\frac{3}{4}$ pound of coffee beans. He uses $\frac{1}{12}$ pound of beans for a pot of coffee. How many pots of coffee can Tito make?

A 8
B 9
C 12
D 16

> This problem asks you to find the pots of coffee that can be made with $\frac{3}{4}$ pound of coffee beans. Divide $\frac{3}{4}$ by $\frac{1}{12}$. To divide fractions, multiply by the reciprocal of the divisor: $\frac{3}{4} \div \frac{1}{12} = \frac{3}{4} \times \frac{12}{1} = \frac{36}{4} = 9$ pots. The correct answer is B.

INDEPENDENT PRACTICE

Read each problem. Circle the letter of the best answer.

1 What is the reciprocal of $\frac{5}{8}$?

A $\frac{5}{8}$
B $\frac{8}{5}$
C $-\frac{5}{8}$
D $-\frac{8}{5}$

2 Of the students in Hillary's class, $\frac{7}{8}$ play sports. Of these students, $\frac{2}{5}$ are girls. What fraction of students in Hillary's class are girls who play sports?

A $\frac{6}{7}$
B $\frac{7}{20}$
C $\frac{19}{40}$
D $\frac{35}{16}$

3 To divide $\frac{5}{9}$ by $\frac{6}{7}$, which two fractions should be multiplied?

A $\frac{5}{9}$ and $\frac{6}{7}$
B $\frac{9}{5}$ and $\frac{6}{7}$
C $\frac{5}{9}$ and $\frac{7}{6}$
D $\frac{9}{5}$ and $\frac{7}{6}$

4 Douglas multiplies $\frac{8}{9}$ by its reciprocal. What number results?

A $-\frac{8}{9}$
B 0
C 1
D $\frac{9}{8}$

5 What is the quotient of $\frac{4}{15}$ and $\frac{2}{3}$?

A $\frac{2}{5}$
B $\frac{4}{5}$
C $\frac{2}{45}$
D $\frac{8}{45}$

6 Paige walked $\frac{9}{10}$ mile. For $\frac{2}{5}$ of the walk, it was raining. What distance did Paige walk in the rain?

A $\frac{1}{2}$ mi
B $\frac{2}{9}$ mi
C $\frac{4}{9}$ mi
D $\frac{9}{25}$ mi

UNIT 3 Operations

GUIDED PRACTICE

Try this sample constructed response problem.

S Of the people responding to a survey, $\frac{5}{8}$ said they watch movies at a theater. Of those people, $\frac{2}{5}$ said they buy popcorn at the theater.

What fraction of the people surveyed buy popcorn when they are at the movie theater?

Show your work.

$$\frac{5}{8} \times \frac{2}{5} = \frac{\cancel{5}^1}{\cancel{8}_4} \times \frac{\cancel{2}^1}{\cancel{5}_1} = \frac{1}{4}$$

Answer: $\frac{1}{4}$

> This problem asks you to find the fraction of people surveyed who buy popcorn when they are at a movie theater. Multiply the fraction of people surveyed who go to the movie theater by the fraction of people who buy popcorn at the theater. The correct answer is $\frac{1}{4}$.

INDEPENDENT PRACTICE

Read the problem. Write your answers.

7 Mr. Johnson has $\frac{5}{6}$ acre of land. He separates the land into sections of $\frac{1}{9}$ acre each.

Part A: Write a multiplication expression that can be used to find how many $\frac{1}{9}$-acre sections of land there are.

Answer: _____

Part B: How many $\frac{1}{9}$-acre sections of land are there?

Answer: _____

UNIT 3 Operations

INDEPENDENT PRACTICE

Read the problem. Write your answers.

8 Look at these fractions.

$$\frac{3}{16} \qquad \frac{9}{12}$$

Part A: What is the value of $\frac{3}{16} \times \frac{9}{12}$? Write your answer in simplest form.

Answer: _____

Part B: What is the value of $\frac{3}{16} \div \frac{9}{12}$? Write your answer in simplest form.

Show your work.

 To divide fractions, use the reciprocal of the divisor.

Answer: _____

UNIT 3 Operations

LESSON 3: Operations with Mixed Numbers

Indicator 6.N.18

mixed numbers   improper fractions

Mixed numbers should be changed into **improper fractions** before they are added or subtracted.

What is $4\frac{1}{2} + 3\frac{4}{5}$?

1. Change each mixed number into an improper fraction:
$4\frac{1}{2} + 3\frac{4}{5} = \frac{9}{2} + \frac{19}{5}$.

2. Rewrite the improper fractions as equivalent fractions with the same denominator. The LCD of the denominators 2 and 5 is 10: $\frac{9}{2} + \frac{19}{5} = \frac{45}{10} + \frac{38}{10}$.

3. Add: $\frac{45}{10} + \frac{38}{10} = \frac{45 + 38}{10} = \frac{83}{10}$.

4. Change the improper fraction to a mixed number: $\frac{83}{10} = 8\frac{3}{10}$.

> An improper fraction has a numerator that is greater than the denominator.

> To change a mixed number to an improper fraction, multiply the denominator by the whole number and add the numerator. This becomes the new numerator. The denominator stays the same.
> $1\frac{3}{4} = \frac{4 \times 1 + 3}{4} = \frac{7}{4}$

Change mixed numbers into improper fractions before multiplying or dividing.

What is $6\frac{2}{3} \times 1\frac{3}{5}$?

1. Rewrite as improper fractions: $6\frac{2}{3} \times 1\frac{3}{5} = \frac{20}{3} \times \frac{8}{5}$.

2. Multiply: $\frac{20}{3} \times \frac{8}{5} = \frac{160}{15}$.

3. Write as a mixed number: $\frac{160}{15} = 10\frac{2}{3}$.

> To change an improper fraction to a mixed number, divide the numerator by the denominator. Write the remainder over the denominator.
> $\frac{7}{4} = 4\overline{)7} = 1\frac{3}{4}$, remainder $\frac{3}{4}$

> You can cancel common factors when multiplying fractions to make the numbers easier to work with.

UNIT 3 Operations

GUIDED PRACTICE

Try this sample multiple-choice problem.

S Mackenzie bought $5\frac{1}{4}$ pounds of watermelon and $1\frac{7}{8}$ pounds of bananas. How much heavier was the watermelon than the bananas?

A $3\frac{3}{8}$ lb

B $3\frac{5}{8}$ lb

C $4\frac{3}{8}$ lb

D $4\frac{5}{8}$ lb

> This problem asks you to find the difference between two weights. First change the mixed numbers to improper fractions: $5\frac{1}{4} - 1\frac{7}{8} = \frac{21}{4} - \frac{15}{8}$. Rewrite the improper fractions as equivalent fractions with the same denominator and subtract: $\frac{21}{4} - \frac{15}{8} = \frac{42}{8} - \frac{15}{8} = \frac{27}{8}$. Divide the numerator by the denominator to write the improper fraction as a mixed number: $\frac{27}{8} = 3\frac{3}{8}$. The correct answer is A.

INDEPENDENT PRACTICE

Read each problem. Circle the letter of the best answer.

1 A bread recipe uses $1\frac{3}{4}$ cups of wheat flour and $1\frac{2}{3}$ cups of rye flour. How much flour is used altogether?

A $2\frac{5}{12}$ cups   C $3\frac{5}{12}$ cups

B $2\frac{5}{7}$ cups   D $3\frac{5}{7}$ cups

2 Toby cut off $\frac{1}{3}$ the length of a $4\frac{5}{6}$-foot board. How long was the piece Toby cut off?

A $1\frac{1}{3}$ ft   C $1\frac{11}{18}$ ft

B $1\frac{5}{18}$ ft   D $1\frac{2}{3}$ ft

3 A bag contains $3\frac{1}{2}$ cups of popcorn kernels. A popcorn maker uses $\frac{3}{8}$ cup of kernels for each batch of popped popcorn. How many batches can be made from this bag of popcorn kernels?

A $1\frac{5}{16}$ C $8\frac{1}{2}$

B $3\frac{7}{8}$ D $9\frac{1}{3}$

4 Yesterday Tamika rode her bike $8\frac{3}{10}$ miles. Today she rode her bike $2\frac{4}{5}$ miles. How much farther did Tamika ride her bike yesterday than today?

A $5\frac{1}{5}$ mi   C $6\frac{1}{5}$ mi

B $5\frac{1}{2}$ mi   D $6\frac{1}{2}$ mi

UNIT 3 Operations

GUIDED PRACTICE

Try this sample constructed response problem.

S Felicity typed an $8\frac{1}{2}$-page book report in 3 hours.

What was the average number of pages she typed each hour?

Show your work.

$$8\frac{1}{2} \div 3\frac{3}{4} = \frac{17}{2} \div \frac{15}{4}$$

$$\frac{17}{\cancel{2}_1} \times \frac{\cancel{4}^2}{15} = \frac{34}{15} = 2\frac{4}{15}$$

Answer: $2\frac{4}{15}$

> This problem asks you to find the average number of pages typed in an hour. Divide the number of pages typed by the number of hours typed: $8\frac{1}{2} \div 3\frac{3}{4}$. Change the mixed numbers to improper fractions. Then, multiply by the reciprocal of the divisor. Finally, change the improper fraction to a mixed number and simplify. The correct answer is $2\frac{4}{15}$.

INDEPENDENT PRACTICE

Read the problem. Write your answers.

5 Stefano walked $1\frac{1}{3}$ hours at an average rate of $2\frac{1}{5}$ miles an hour.

How far did Stefano walk?

Show your work.

Answer: _____

UNIT 3 Operations

Read the problem. Write your answers.

6 Ron and Jolanda both worked last weekend.

- Ron worked $5\frac{3}{4}$ hours Saturday and $3\frac{1}{2}$ hours Sunday.

- Jolanda worked $8\frac{2}{3}$ hours Saturday and 6 hours Sunday.

Part A: How many hours did Ron work last weekend?

Show your work.

Answer: _____

Part B: How many more hours did Jolanda work last weekend than Ron?

Show your work.

> Find the total number of hours that Jolanda worked.

Answer: _____

Exponential Forms

Indicators 6.N.23, 24, 25

exponent   base   power   exponential form

An **exponent** shows the number of times a number, called the **base**, is multiplied by itself. This product is known as the **power.**

$n^x = p$ $10^2 = 100$
n to the x power is p.   10 to the 2nd power is 100.

In the number 10^2, the base is 10, the exponent is 2, and the power is 100.

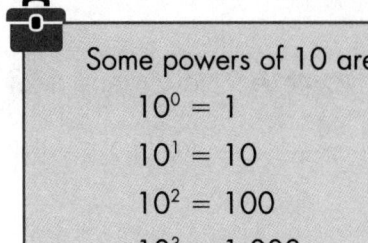

Some powers of 10 are:
$10^0 = 1$
$10^1 = 10$
$10^2 = 100$
$10^3 = 1,000$

What is the value of 5^3?

1. Write the base, 5, multiplied by itself 3 times: $5^3 = 5 \times 5 \times 5$.

2. Multiply: $5 \times 5 \times 5 = 25 \times 5 = 125$. The value of 5^3 is 125.

A number multiplied by itself repeatedly can be written in **exponential form.**

How is $7 \times 7 \times 7 \times 7 \times 7$ written in exponential form?

1. Identify the base, the number that is being multiplied: 7.

2. Identify the exponent. Count the number of times the base number is multiplied by itself: 7 is multiplied by itself 5 times. So the exponent is 5.

3. Write the number in exponential form: 7^5.

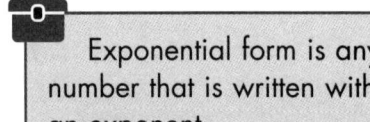

Exponential form is any number that is written with an exponent.

2^3 is the exponential form of
$2 \times 2 \times 2$ and 8.

Any number with an exponent of 1 equals that number.

$15^1 = 15$ $(-6)^1 = -6$ $0^1 = 0$

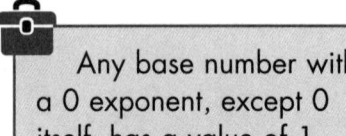

Any base number with a 0 exponent, except 0 itself, has a value of 1.

$8^0 = 1$ $(-\frac{1}{4})^0 = 1$

UNIT 3 Operations

GUIDED PRACTICE

Try this sample multiple-choice problem.

S The expression s^2 represents the area of a square with side length s. What is the area of a square with a side length of 6 inches?

- **A** 8 sq in.
- **B** 12 sq in.
- **C** 36 sq in.
- **D** 62 sq in.

> This problem asks you to find the value of s^2 when $s = 6$ inches. To do this, substitute 6 for s: 6^2. The exponent 2 means multiply the base 6 by itself two times: $6^2 = 6 \times 6 = 36$. The correct answer is C.

INDEPENDENT PRACTICE

Read each problem. Circle the letter of the best answer.

1 Which expression is equivalent to 25^2?

- **A** 2^{25}
- **B** 25×2
- **C** $25 + 25$
- **D** 25×25

2 Which expression is equal to 1?

- **A** 0^1
- **B** $5{,}000^0$
- **C** $5^1 \times 1^5$
- **D** $1 \times 1 \times 1 \times 0$

3 Which shows another way of writing this expression?

$$10 \times 10 \times 10 \times 10$$

- **A** 10^4
- **B** 40^4
- **C** 10×4
- **D** 40,000

4 What is the value of 4^3?

- **A** 12
- **B** 43
- **C** 64
- **D** 81

5 Which expression has the same value as 2^7?

- **A** 2×7
- **B** 7×7
- **C** $2 \times 7 \times 2 \times 7$
- **D** $2 \times 2 \times 2 \times 2 \times 2 \times 2 \times 2$

6 Bryan raised a base number 8 to the power of 5. Which expression shows this?

- **A** 5^8
- **B** 8×5
- **C** $8 \times 8 \times 8 \times 8 \times 8$
- **D** $5 \times 5 \times 5 \times 5 \times 5 \times 5 \times 5 \times 5$

UNIT 3 Operations

GUIDED PRACTICE

Try this sample constructed response problem.

S Look at this number written in exponential form.

$$6^4$$

Part A: What number is the base? What number is the exponent?

Base: _____6_____

Exponent: _____4_____

Part B: How is this number written as a multiplication expression?

Answer: ____6 × 6 × 6 × 6____

> Part A asks you to identify the base and the exponent in 6^4. The base is the number being multiplied by itself. The exponent tells the number of times the base is multiplied. In this number, 6 is the base and 4 is the exponent. Part B asks you to write a multiplication expression that is equivalent to 6^4. The expression should show the base 6 multiplied by itself 4 times. The correct answer is 6 × 6 × 6 × 6.

INDEPENDENT PRACTICE

Read the problem. Write your answers.

7 Chad wrote the expression 3 × 3 × 3 × 3 × 3 × 3.

   Part A: How is this number written in exponential form?

   Answer: _____

   Part B: What number represents the exponent in part A?

   Answer: _____

66 UNIT 3 Operations

INDEPENDENT PRACTICE

Read the problem. Write your answers.

8 The number 9^3 represents the volume of a cube with a side length of 9 centimeters.

Part A: What number is the base? What number is the exponent?

Base: _____

Exponent: _____

Part B: What is the volume, in cubic centimeters, of this cube? Show your work.

> Write 9^3 as a multiplication expression.

Answer: _____

Order of Operations

Indicators 6.N.22, 25

order of operations   parentheses   exponents   multiply
divide   add   subtract

Some expressions contain more than one operation.

$6 \div 2 + 7 \quad 6 \times 2 - 7 \quad 6 - 2 + 7$

The values of these expressions are all different.

When an expression has more than one operation, use the **order of operations** to find its value.

What is the value of this expression?

$16 + 2^3 \div (5 - 1) \times 2$

1. Perform the operations inside **parentheses.**

    $16 + 2^3 \div (5 - 1) \times 2$
    $16 + 2^3 \div 4 \times 2$

2. Evaluate **exponents.**

    $16 + 2^3 \div 4 \times 2$
    $16 + 8 \div 4 \times 2$

3. **Multiply** or **divide** in order from left to right.

    $16 + 8 \div 4 \times 2$
    Division is first: $16 + 2 \times 2$
    Multiplication is next: $16 + 4$

4. **Add** or **subtract** in order from left to right.

    $16 + 4$
    20

Parentheses ( ) are grouping symbols. **Always** work inside parentheses first.

If there are no parentheses, follow the rest of the steps in order.

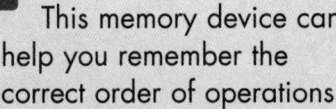

This memory device can help you remember the correct order of operations:

Please = **P**arentheses
Excuse = **E**xponents
My   = **M**ultiplication
Dear   = **D**ivision
Aunt   = **A**ddition
Sally = **S**ubtraction

The letters *PEMDAS* can also be used as a reminder.

UNIT 3 Operations

GUIDED PRACTICE

Try this sample multiple-choice problem.

S What is the value of the expression below?

$$6 + 24 \div 2^3 \times 2$$

- **A** 8
- **B** 10
- **C** 12
- **D** 14

> This problem asks you to find the value of the expression. Follow the correct order of operations. Since there are no parentheses, start with the exponent: $6 + 24 \div 2^3 \times 2 = 6 + 24 \div 8 \times 2 = 6 + 3 \times 2 = 6 + 6 = 12$. The correct answer is C.

INDEPENDENT PRACTICE

Read each problem. Circle the letter of the best answer.

1 Which step should be performed first when evaluating the expression below?

$$8 \div 4 + (9 - 6) \div 3$$

- **A** $8 \div 4$
- **B** $4 + 9$
- **C** $9 - 6$
- **D** $6 \div 3$

2 Look at this expression.

$$(5 - 1)^3 \times 2$$

What is the value of the expression?

- **A** 3
- **B** 8
- **C** 24
- **D** 128

3 Which statement about this expression is true?

$$54 \div (24 - 6) \times 3 + 21$$

- **A** Subtraction is done before division.
- **B** Addition is done before subtraction.
- **C** Multiplication is done before division.
- **D** Addition is done before multiplication.

4 Which expression has the smallest value?

- **A** $(15 + 6) \div 3 - 1$
- **B** $15 + (6 \div 3) - 1$
- **C** $15 + 6 \div (3 - 1)$
- **D** $15 + 6 \div 3 - 1$

5 In which order should you perform the operations to find the value of this expression?

$$20 - (30 \div 5 \times 2) + 10$$

- **A** multiply, divide, add, subtract
- **B** divide, multiply, subtract, add
- **C** multiply, divide, subtract, add
- **D** subtract, divide, multiply, add

6 What is the value of this expression?

$$2 \times (6^2 \div 3) - 5 + 1$$

- **A** 2
- **B** 4
- **C** 18
- **D** 20

UNIT 3 Operations

GUIDED PRACTICE

Try this sample constructed response problem.

S Look at this expression.

$$5^2 - (4 + 2) \div 3 \times 6$$

Part A: In what order should the four basic operations be performed when evaluating this expression?

Answer: <u>addition, division,</u>

<u>multiplication, subtraction</u>

Part B: What is the value of this expression?

Answer: <u>   13   </u>

> Part A asks you to identify the correct order of operations. Always work inside parentheses first. So addition is first. Multiply and divide from left to right. So division is next, then multiplication. Finally, subtraction is performed. The correct answer is addition, division, multiplication, and subtraction. Part B asks you to find the value of the expression:
> $5^2 - (4 + 2) \div 3 \times 6 =$
> $5^2 - 6 \div 3 \times 6 =$
> $25 - 6 \div 3 \times 6 =$
> $25 - 2 \times 6 = 25 - 12 = 13.$
> The correct answer is 13.

INDEPENDENT PRACTICE

Read the problem. Write your answers.

7 Look at this expression.

$$1 + 8 \times (6 - 4)^3$$

What is the value of this expression?

Show your work.

Answer: _____

Read the problem. Write your answers.

8 Look at this expression.

$$7 + 3 \times (4^2 - 6)$$

Part A: What is the value of this expression?

Show your work.

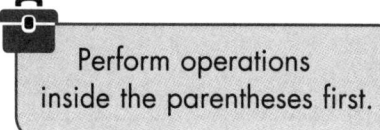

Perform operations inside the parentheses first.

Answer: _____

Part B: How can you move the parentheses in this expression so that its value becomes 154?

Answer: _____

Lesson 6: Estimation with Percents

Indicators 6.N.26, 27

estimation   round   reasonable

Estimation is used to get an approximate answer when an exact answer is not needed. Percents can be rounded to the nearest 1% or nearest 10% to make estimation easier.

A restaurant charges an 18% fee for delivering all food orders. **About** how much can a family expect to pay for delivery on a food order costing $36?

1. Round 18% to the nearest 10%: 18% rounds up to 20%.

2. Find 20% of $36: 20% × $36 = 0.2 × $36 = $7.20. The delivery fee is **about** $7.20.

> To **round** a number, look at the digit to the right of the place you are rounding to. If it is 4 or lower, round down. If it is 5 or higher, round up.

> Some percents will be rounded to the nearest 1%. Others will be rounded to the nearest 10%.
> 16.8% rounds up to 17%.
> 81% rounds down to 80%.

You can use estimation to check that the answer to a computation is **reasonable**.

A research company mailed questionnaires to 408 families. Only 32% of the families responded. An analyst calculates that 94 families responded. Is this answer reasonable?

1. Round 408 to the nearest hundred: 408 rounds to 400.

2. Round 32% to the nearest 10%: 32% rounds to 30%.

3. Find 30% of 400: 30% × 400 = 0.3 × 400 = 120.

4. Compare the estimate to the calculation. The number of families that responded is **about** 120. So, the analyst's calculation is **not** reasonable.

> Round only numbers with more than one digit.

72 **UNIT 3** Operations

© The Continental Press, Inc.   Do not duplicate.

GUIDED PRACTICE

Try this sample multiple-choice problem.

S Missouri charges a sales tax of 4.225%. Jack ordered some books totaling $45.88 from a company in Missouri and has to pay sales tax. **About** how much will Jack pay in sales tax?

- A $1.84
- B $2.25
- C $3.75
- D $4.23

> This problem asks you to estimate the amount of sales tax on an order totaling $45.88. To do this, first round each amount: $45.88 rounds up to $46 and 4.225% rounds down to 4%. Next, multiply the rounded amounts: $46 × 4% = $46 × 0.04 = $1.84. The correct answer is A.

INDEPENDENT PRACTICE

Read each problem. Circle the letter of the best answer.

1 What is the **best** estimate of 77% of 8.9?

- A 5.6
- B 6.0
- C 7.2
- D 7.7

2 Julio answered 82% of the 58 questions on a test correctly. Which calculation is **best** for Julio to use when estimating the number of test questions he answered correctly?

- A 8 × 60
- B 0.8 × 60
- C 9 × 50
- D 0.9 × 50

3 Lydia used her calculator to compute 14.375% of 6,172. The screen on her calculator showed the answer as 86.408. Is this answer reasonable?

- A Yes, the answer should be **about** 86.
- B No, the answer should be **about** 286.
- C No, the answer should be **about** 600.
- D No, the answer should be **about** 840.

4 A file is 484 kilobytes in size. Only 64% of it downloaded before stopping. **About** how many kilobytes downloaded?

- A 300
- B 350
- C 400
- D 450

5 A box of cereal claims that it has 27% more cereal than a box with 11.2 ounces. **About** how many more ounces are in the box of cereal?

- A 2.5
- B 3.3
- C 3.9
- D 4.5

6 The regular price of a winter coat is $187. It is on sale for 33% off. Bethany calculates the amount of her savings is $61.71. Is her answer reasonable?

- A Yes, the answer should be **about** $60.
- B No, the answer should be **about** $40.
- C No, the answer should be **about** $80.
- D No, the answer should be **about** $100.

UNIT 3 Operations

GUIDED PRACTICE

Try this sample constructed response problem.

S Sasha set her budget to have 36% of her earnings put into a savings account. Last week, she earned $521.93. She wants to estimate the amount of last week's earnings that will be put into her savings account.

Part A: What numbers should Sasha round to when she estimates the amount that will be put into her savings account?

Earnings round to: ___$500___

Percent rounds to: ___40%___

Part B: **About** how much will Sasha put into her savings account?

Answer: ___$200___

> Part A asks you to round the earnings amount and the percent used for estimating. The amount $521.93 rounded to the nearest hundred is $500. The percent 36% rounded to the nearest 10% is 40%. Part B asks you to find the estimated amount. Multiply the rounded dollar amount by the rounded percent: $500 × 40% = $500 × 0.4 = $200. The correct answer is $200.

INDEPENDENT PRACTICE

Read the problem. Write your answers.

7 Andrew read a report saying that the population of his town increased by 8.2% over the past year. The report said that last year the town population was 31,589. Andrew calculated the increase in population as 2,590 people.

Is Andrew's calculation reasonable?

Answer: _____

Use estimation to justify your answer.

_____
_____
_____
_____

UNIT 3 Operations

INDEPENDENT PRACTICE

Read the problem. Write your answers.

8 A low-fat snack has 15.4% fewer calories than the original snack. The original snack has 195 calories in each serving.

Part A: Write a multiplication expression that can be used to estimate the number of total calories in each serving of the low-fat snack.

Think about how finding the number of total calories in the low-fat snack is different from finding the number of fewer calories in the low-fat snack.

Answer: _____

Explain why the numbers in your expression are reasonable.

_____

_____

_____

_____

Part B: About how many total calories are in each serving of the low-fat snack?

Answer: _____

UNIT 3 Operations

Operations Review

Read each problem. Circle the letter of the best answer.

1 How is 11 × 11 × 11 × 11 × 11 written in exponential form?

  A 11 × 5   **C** 11^5

  B 5^{11} **D** 55^5

2 Which expression can be used to find the quotient of $\frac{2}{7}$ and $\frac{7}{10}$?

  A $\frac{2}{7} \times \frac{7}{10}$

  B $\frac{7}{2} \times \frac{10}{7}$

  C $\frac{7}{2} \times \frac{7}{10}$

  D $\frac{2}{7} \times \frac{10}{7}$

3 Which expression has the same value as 7^3?

  A 73

  B 7 × 3

  C 7 × 7 × 7

  D 3 × 3 × 3 × 3 × 3 × 3 × 3

4 A box contains geometric shapes. Squares make up $\frac{3}{5}$ of the shapes. Of the squares, $\frac{5}{6}$ are yellow. What fraction of the shapes in the box are yellow squares?

  A $\frac{1}{2}$ **C** $\frac{7}{30}$

  B $\frac{8}{11}$ **D** $1\frac{13}{30}$

5 Which expression has the greatest value?

  A 3^3 **C** 8^1

  B 4^2 **D** 9^0

6 Kyle mailed an envelope that weighed $\frac{7}{16}$ pound. Hank mailed an envelope that weighed $\frac{3}{4}$ pound. How much heavier was the envelope Hank mailed than the envelope Kyle mailed?

  A $\frac{1}{4}$ pound

  B $\frac{5}{16}$ pound

  C $\frac{1}{3}$ pound

  D $\frac{3}{8}$ pound

7 A 15.8-ounce jar of mixed nuts says that it contains 39% peanuts. **About** how many ounces of the mixed nuts are peanuts?

  A 5.3   **C** 7.6

  B 6.4   **D** 8.0

8 A box contains $12\frac{1}{4}$ servings of cereal. Each serving is $1\frac{1}{3}$ cups. How many total cups of cereal are in a full box?

  A $9\frac{3}{16}$ **C** $13\frac{7}{12}$

  B $12\frac{1}{12}$ **D** $16\frac{1}{3}$

Operations Review

Read each problem. Write your answers.

9 Hanging plants at a farm stand cost $17.69 each. If four or more plants are bought, each plant's price is reduced by 33%. Ms. Carnevale decides to buy 5 hanging plants. She calculates that she will save a total of $29.20 altogether.

Is her calculation reasonable?

Answer: _____

Use estimation to justify your answer.

_____
_____
_____
_____

10 Look at this numerical expression.

$$6 + 16 \div 2 \times (7 - 5)$$

Part A: In what order should the operations be performed to evaluate this expression?

Answer: _____

Part B: What is the value of this expression?

Answer: _____

UNIT 3 Operations

Operations Review

Read the problem. Write your answers.

11 A soup recipe uses $2\frac{1}{2}$ cups of chicken broth and $1\frac{7}{8}$ cups of vegetable broth.

Part A: How many total cups of broth does this recipe use?

Answer: _____

How many more cups of chicken broth does this recipe use than vegetable broth?

Answer: _____

Part B: Emma has only $1\frac{3}{4}$ cups of chicken broth. What fraction of the recipe can she make with the chicken broth she has?

Show your work.

Answer: _____

Algebra

When you use a letter to stand for a number you do not know, you are using algebra. Algebra is a way to make general statements about mathematical relationships. You can show these relationships using expressions and equations. Have you ever used a formula to find the perimeter of a rectangle or the amount of time it will take to travel a certain distance? Formulas also show mathematical relationships. When you solve a formula, you are using algebra.

This unit will help you answer test questions about algebra. There are four lessons in this unit:

1. **Variables and Expressions** This lesson reviews how to write algebraic expressions that describe a given situation.

2. **Evaluating Expressions** This lesson reviews how to find the value of an algebraic expression when you are given the value of the variable.

3. **Solving Equations** In this lesson, you will review how to solve an equation using inverse operations. You will also check your solution to find out if it is right.

4. **Evaluating Formulas** This lesson reviews how to use formulas for circumference, perimeter, area, volume, distance, interest, and temperature.

Variables and Expressions

Indicators 5.A.2; 6.A.1

algebraic expression   variable   verbal expression

An **algebraic expression** is an expression containing a **variable,** or letter that stands for a number.

$$s + 2 \quad y \div 3 \quad 4x + 1$$

 A variable can be any letter, such as $a, b, d, n, x,$ or y.

A **verbal expression** is an expression that uses words. It can be translated into an algebraic expression.

Write an algebraic expression that shows this verbal expression: eight more than a number.

1. Identify the number and variable from the verbal expression. The number is 8. Let the variable be n.

2. Decide how the number and variable are to be combined. The words *more than* in this expression mean addition. The number 8 should be added to n. The expression is $n + 8$.

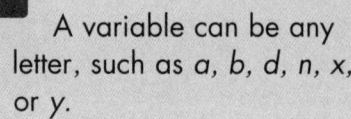 Some words give clues about operations.
Total, in all, more than, or *altogether* may indicate addition or multiplication.
Difference, less than, fewer, or *left over,* may indicate subtraction.
Divided among, shared, or *split equally* may indicate division.

Some expressions involve more than one step.

What algebraic expression names six less than three times a number?

1. Identify the numbers and variable from the verbal expression. The numbers are 6 and 3. Let the variable be n.

2. Decide how the numbers and variable are related. Three times a number means $3n$. The words *less than* in this expression mean subtraction. So, 6 should be subtracted from $3n$. The expression is $3n - 6$.

 The variable x can be confused for the multiplication symbol $\times$. Use a different letter for the variable if the expression involves multiplication.
$$5y = 5 \times y$$

UNIT 4 Algebra

© The Continental Press, Inc.   Do not duplicate.

GUIDED PRACTICE

Try this sample multiple-choice problem.

S The length of a rectangle is 10 more than half its width, w. Which expression represents the length of the rectangle?

A $\frac{1}{2}(10) + w$ C $\frac{1}{2}(w + 10)$

B $\frac{1}{2}w + 10$ D $\frac{1}{2} + w + 10$

> This problem asks you to identify the algebraic expression that describes the length of a rectangle. This length is 10 more than half the width. Half the width means multiply: $\frac{1}{2}w$. Ten more than that means add 10. The algebraic expression is $\frac{1}{2}w + 10$. The correct answer is B.

INDEPENDENT PRACTICE

Read each problem. Circle the letter of the best answer.

1 Which expression means the quotient of p and 4?

A $4p$ C $p - 4$

B $4 + p$ D $\frac{p}{4}$

2 Padma exercised 3 days this week for the same number of minutes each day. Which expression shows the total number of minutes she exercised this week?

A $3 + m$ C $3m$

B $m - 3$ D $\frac{m}{3}$

3 Five friends decide to share the cost of a DVD player and some DVDs. The DVD player costs $240. Each DVD costs $14. Which expression shows the total amount each friend will pay towards the total cost of the DVD player and d DVDs?

A $\frac{d(240 + 14)}{5}$ C $\frac{240 + 14d}{5}$

B $5(240 + 14d)$ D $5(d + 240 + 14)$

4 Jenna has b fewer books than her sister. Her sister has 35 books. Which expression shows the number of books Jenna has?

A $b - 35$ C $35 + b$

B $35 - b$ D $b \div 35$

5 An adult concert ticket costs $15 less than twice a child's ticket. Which expression represents the price of an adult ticket?

A $2c - 15$ C $15 - 2c$

B $2(c - 15)$ D $2(15 - c)$

6 Max is x inches taller this summer than last summer. Last summer, he was 58 inches tall. Which expression can be used to find Max's height **in feet** this summer?

A $12(58 - x)$ C $\frac{58 - x}{12}$

B $12(58 + x)$ D $\frac{58 + x}{12}$

UNIT 4 Algebra

GUIDED PRACTICE

Try this sample constructed response problem.

S Jermaine gets paid $9 per hour to deliver food.

Part A: Write an expression that shows how much Jermaine earns working h hours.

Answer: ____9h____

Part B: Jermaine also earns $0.50 per mile he drives. One week, he worked 10 hours and drove m miles. Write an expression to show how much money Jermaine earned that week.

Answer: ____90 + 0.50m____

> Part A asks you to write an expression that shows how much is earned in h hours at $9 per hour. The words *per hour* tell you to multiply. The correct answer is $9h$. Part B asks you to write an expression that shows the total amount that is earned driving m miles. The amount paid per mile is $0.50, so the expression $0.50m$ shows the amount paid for m miles. This is added to the regular pay of $90. The correct answer is $90 + 0.50m$.

INDEPENDENT PRACTICE

Read the problem. Write your answers.

7 Nina is eight years younger than Sophie.

Part A: Write an expression to show Nina's age if Sophie is y years old.

Answer: _____

Part B: In five years, Nina will be half of Sophie's age. Write an expression to show Nina's age five years from now.

Answer: _____

INDEPENDENT PRACTICE

Read the problem. Write your answers.

8 Kelly wants to mail two packages. The first package weighs n ounces.

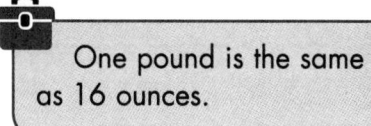
One pound is the same as 16 ounces.

Part A: Write an expression to show the weight of the first package, in pounds.

Answer: _____

Part B: The second package weighs one pound more than twice the weight of the first package. Write an expression for the weight of the second package, in pounds.

Answer: _____

Write an expression for the combined weight, in pounds, of both packages.

Answer: _____

Lesson 2: Evaluating Expressions

Indicators 5.A.3; 6.A.2

evaluate   substitute   exponent   base

To **evaluate** an expression means to find its value. To do this, **substitute** a given value for the variable and carry out the operation.

What is the value of $q + 16$ when $q = 37$?

1. Substitute 37 for q: $37 + 16$.

2. Add: $37 + 16 = 53$. The value of $q + 16$ when $q = 37$ is 53.

Some expressions have more than one operation. To evaluate these, substitute the given value for the variable. Then simplify the expression using the correct order of operations.

What is the value of $\frac{z-5}{4}$ when $z = 11$?

1. Substitute 11 for z: $\frac{11-5}{4}$.

2. Simplify using the correct order of operations.

   First subtract: $\frac{11-5}{4} = \frac{6}{4}$.

   Then simplify: $\frac{6}{4} = \frac{3}{2}$.

   The value of $\frac{z-5}{4}$ when $z = 11$ is $\frac{3}{2}$ or $1\frac{1}{2}$.

The order of operations is:

Parentheses

Exponents

Multiplication and **D**ivision from left to right

Addition and **S**ubtraction from left to right

An **exponent** is a number that tells how many times another number, called the **base,** multiplies itself:

Exponent
↓
$5^2 = 5 \times 5 = 25$
↑
Base

UNIT 4 Algebra

GUIDED PRACTICE

Try this sample multiple-choice problem.

S What is the value of $k^3 - 9$ when $k = 4$?

    A 3

    B 7

    C 55

    D 125

> This problem asks you to find the value of $k^3 - 9$ given that $k = 4$. Substitute 4 for k into the expression: $4^3 - 9$. Then simplify using the correct order of operations. Evaluate exponents before subtracting: $4^3 - 9 = (4 \times 4 \times 4) - 9 = 64 - 9 = 55$. The correct answer is C.

INDEPENDENT PRACTICE

Read each problem. Circle the letter of the best answer.

1 What is the value of $j - 7$ when j is 16?

    A 9   C 21

    B 11   D 23

2 The expression below shows the cost, in dollars, of a cooler and b bags of ice.

$$45 + 2b$$

What is the total cost of the cooler and 12 bags of ice?

    A $49   C $69

    B $59   D $79

3 The volume of a cube with a side length of s is s^3 cubic units. What is the volume of a cube with a side length of 6 centimeters?

    A 9 cm³

    B 18 cm³

    C 63 cm³

    D 216 cm³

4 Evaluate the expression $(5 + c)^2$ when $c = 4$.

    A 13   C 21

    B 18   D 81

5 The actual time, in minutes, of a radio program with c commercials is represented by the expression below.

$$30 - \frac{1}{4}c$$

What is the length of a radio show with 40 commercials?

    A 10 minutes   C 20 minutes

    B 12 minutes   D 24 minutes

6 A square garden measures 7 feet on each side. Pedro plans to increase the length of each side of the garden by x feet. The area of the new garden, in square feet, is represented by $(7 + x)^2$. What is the area of the new garden if $x = 3$?

    A 16 sq ft   C 160 sq ft

    B 100 sq ft   D 441 sq ft

GUIDED PRACTICE

Try this sample constructed response problem.

S At a store, each loaf of bread costs the same amount. Iliana uses the expression below to determine her total cost of 4 loaves of sourdough bread and 1 loaf of pumpernickel bread.

$$(4 + 1)d$$

What is Iliana's total cost if $d = \$2.50$?

Answer: ___$12.50___

> This problem asks you to find the value of $(4 + 1)d$ when $d = \$2.50$. First substitute 2.50 for d: $(4 + 1)(2.50)$. Next, simplify the expression using the correct order of operations. Parentheses are evaluated first. Then multiplication is performed next: $(4 + 1)(2.50) = 5(2.50) = 12.50$. The correct answer is $12.50.

Explain how you found your answer.

INDEPENDENT PRACTICE

Read the problem. Write your answers.

7 Chet uses the expression below to find the cost to rent roller skates and then skate for h hours.

$$6 + 4.50h$$

What is the total cost, in dollars, to skate for 2 hours?

Answer: _____

Explain how you know your answer is correct.

_____
_____
_____
_____

86   **UNIT 4** Algebra

INDEPENDENT PRACTICE

Read the problem. Write your answers.

8 Look at the expression below.

$$\frac{(16 - m)^2 + 6}{2}$$

Part A: What is the value of the expression when $m = 4$?

Show your work.

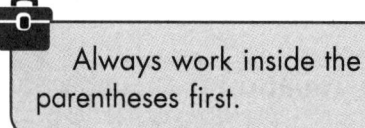

Always work inside the parentheses first.

Answer: _____

Part B: What is the value of the expression when the parentheses are removed?

Show your work.

Answer: _____

Lesson 3: Solving Equations

Indicators 5.A.4, 5

equation   algebraic equations   solve   inverse operations   isolate

An **equation** is a number sentence that shows two expressions are equal. Both sides of an equation must be balanced in order for the equation to be true.

The equations below are true.

$$11 - 2 = 9 \qquad 5 \times 3 = 8 + 7$$
$$9 = 9 \qquad\qquad 15 = 15$$

The equations below are **not** true.

$$4 + 7 \stackrel{?}{=} 28 \qquad 12 \div 3 \stackrel{?}{=} 2 \times 3$$
$$11 \neq 28 \qquad\qquad 4 \neq 6$$

The symbol $\neq$ means "is not equal to."

Some equations contain just numbers. Other equations contain numbers and a variable. These are known as **algebraic equations.**

You can **solve** an equation to find the value of a variable using **inverse operations.** These "undo" a given operation. The correct value of the variable is the value that makes the equation true.

What is the value of t in the equation below?

$$t \div 3 = 12$$

1. **Isolate,** or set apart, the variable by using the inverse operation. Multiplication is the inverse, or opposite, of division: $t \div 3 \times 3 = 12 \times 3 \rightarrow t = 12 \times 3$.

2. Simplify: $t = 12 \times 3 = 36$.

3. Check the solution to an equation by substituting the answer for the variable into the original equation.

   Check: $t \div 3 = 12 \rightarrow 36 \div 3 = 12$
   $12 = 12$, so $t = 36$

Inverse operations are opposite operations.
   Addition and subtraction are inverse operations.
   Multiplication and division are inverse operations.

To keep an equation in balance, the same operation is performed on **both** sides.

UNIT 4 Algebra

GUIDED PRACTICE

Try this sample multiple-choice problem.

S Brent spent $14 on a gift for a friend and had $39 left. The equation $d - 14 = 39$ can be used to find the amount of money Brent had before he bought the gift. What is the value, in dollars, of d?

A 23

B 25

C 53

D 55

> This problem asks you to find the value of d in the equation $d - 14 = 39$. First isolate the variable. Use the inverse operation of subtraction on both sides of the equation. Addition is the inverse of subtraction, so add 14 to both sides: $d - 14 + 14 = 39 + 14$. This gives $d = 39 + 14 = 53$. The correct answer is C.

INDEPENDENT PRACTICE

Read each problem. Circle the letter of the best answer.

1 What is the value of f in the equation $2f = 6$?

A 3 C 8

B 4 D 12

2 What operation should be performed on each side of the equation below to solve for w?

$$w + 7 = 21$$

A add 7 C add 21

B subtract 7 D subtract 21

3 Adrien paid a total of $22 for a food bowl and some dog food. The dog food cost $13. The equation $13 + b = 22$ can be used to find the cost, in dollars, of the food bowl. How much did the food bowl cost?

A $7 C $11

B $9 D $35

4 In which equation is the variable isolated by dividing each side by 10?

A $10x = 50$ C $x - 10 = 50$

B $10 + x = 50$ D $x \div 10 = 50$

5 What is the solution to $u - 53 = 27$?

A 26 C 70

B 34 D 80

6 Look at this equation.

$$y \times 3 = 12$$

Which equation shows how to solve for y?

A $y \times 3 + 3 = 12 + 3$

B $y \times 3 - 3 = 12 - 3$

C $y \times 3 \times 3 = 12 \times 3$

D $y \times 3 \div 3 = 12 \div 3$

GUIDED PRACTICE

Try this sample constructed response problem.

S The equation below can be used to find the approximate length of the radius of a wheel with a circumference of 30 inches.

$$6 \times r = 30$$

Part A: What operation is performed on both sides of the equation to solve for r?

Answer: _____division_____

Part B: What is the value of r?

Answer: _____5 inches_____

> Part A asks you to identify the operation used to find the value of r in the equation $6 \times r = 30$. The inverse operation is used to solve for r. The inverse of multiplication is division. The correct answer is division. Part B asks you to solve the equation for r. Divide both sides of the equation by 6: $6 \times r \div 6 = 30 \div 6$. So $r = 30 \div 6 = 5$. The correct answer is 5 inches.

INDEPENDENT PRACTICE

Read the problem. Write your answers.

7 Look at this equation.

$$n - 126 = 482$$

What value of n makes this equation true?

Answer: _____

Explain how you know your answer is correct.

_____
_____
_____
_____

UNIT 4 Algebra

INDEPENDENT PRACTICE

Read the problem. Write your answers.

8 The area, in square feet, of a rectangle with a length of 15 feet and a width of w feet is shown in the equation below.

$$135 = 15w$$

Part A: What operation is used in this equation?

Answer: _____

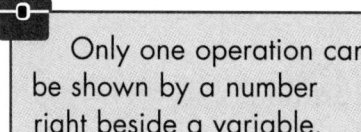

Only one operation can be shown by a number right beside a variable.

What operation should be used to solve for w in this equation?

Answer: _____

Part B: What is the value, in inches, of w?

Show your work.

Answer: _____

UNIT 4 Algebra

Lesson 4: Evaluating Formulas

Indicator 6.A.6

formula circumference perimeter area volume
distance interest temperature evaluate

A **formula** is an important mathematical relationship that is written as an algebraic equation. Some formulas are listed below.

Circumference of a circle	$C = 2\pi r = 2 \times 3.14 \times radius$
Perimeter of a rectangle	$P = 2l + 2w = 2 \times length + 2 \times width$
Perimeter of a square	$P = 4s = 4 \times side$
Area of a rectangle	$A = lw = length \times width$
Area of a triangle	$A = \frac{1}{2}bh = \frac{1}{2} \times base \times height$
Volume of a rectangular prism	$V = lwh = length \times width \times height$
Distance	$d = rt = rate \times time$
Interest	$I = prt = principal \times rate \times time$

> Circumference is the distance around a circle.
> Perimeter is the distance around any shape.

> Area is the amount of space inside a region.

> **Temperature** formulas for Fahrenheit and Celsius degrees:
> $F = \frac{9}{5}C + 32$
> $C = \frac{5}{9}(F - 32)$

You can **evaluate** a formula by substituting given values.

A rectangular prism has a length of 5 meters, a width of 7 meters, and a height of 2 meters. What is the volume, in cubic meters, of this prism?

1. Substitute $l = 5$, $w = 7$, and $h = 2$ into the volume formula: $V = 5 \times 7 \times 2$.

2. Simplify. Use the correct order of operations if needed: $V = 5 \times 7 \times 2 = 70$. The volume of the prism is 70 cubic meters.

> Use the correct order of operations to evaluate formulas:
> **P**arentheses
> **E**xponents
> **M**ultiplication and
> **D**ivision from left to right
> **A**ddition and
> **S**ubtraction from left to right

UNIT 4 Algebra

GUIDED PRACTICE

Try this sample multiple-choice problem.

S John put $500 into a savings account for 1 year. In that year, the money in the savings account earned $10 in interest. What is the interest rate on John's savings account?

- A 1%
- B 2%
- C 5%
- D 10%

> This problem asks you to find the interest rate, r, on a savings account. Solve the interest formula, $I = prt$, for r. Use the inverse operation of multiplication to isolate r: $r = \frac{I}{pt}$. Substitute 10 for I, 500 for p, and 1 for t, and solve: $r = \frac{10}{(500)(1)} = 0.02 = 2\%$. The correct answer is B.

INDEPENDENT PRACTICE

Read each problem. Circle the letter of the best answer.

1 Each side of a square is 80 millimeters long. What is the perimeter of the square?

- A 160 mm
- B 320 mm
- C 640 mm
- D 6,400 mm

2 Carmen drove at a constant speed of 60 miles per hour for $\frac{3}{4}$ hour. How far did she drive in this time?

- A 30 miles
- B 40 miles
- C 45 miles
- D 80 miles

3 The area of a rectangle is 96 square yards. The width of the rectangle is 16 yards. What is the length of the rectangle?

- A 6 yards
- B 8 yards
- C 16 yards
- D 80 yards

4 The circumference of a wagon wheel is 72 centimeters. Which equation can be used to find the approximate length, in centimeters, of the radius of the wheel?

- A $r = 2(3.14)(72)$
- B $r = \frac{72(2)}{3.14}$
- C $r = \frac{72(3.14)}{2}$
- D $r = \frac{72}{2(3.14)}$

5 The base of a triangle is 20 inches. The height of the triangle is 5 inches. What is the area of the triangle?

- A 4 sq in.
- B 8 sq in.
- C 50 sq in.
- D 100 sq in.

6 One week, the temperature of the water in the ocean was 68° Fahrenheit. What is this temperature in degrees Celsius?

- A 20°C
- B 56°C
- C 90°C
- D 154 °C

UNIT 4 Algebra

GUIDED PRACTICE

Try this sample constructed response problem.

S The base of a rectangular box is 3 feet long and 2 feet wide.

   Part A: What is the perimeter, in feet, of the base of the box?

   Answer: __10 feet__

   Part B: The box is 2 feet tall. What is the volume, in cubic feet, of the box?

   Answer: __12 cubic feet__

> Part A asks you to find the perimeter of a rectangle. The formula for the perimeter of a rectangle is $P = 2l + 2w$. Substitute 3 for l and 2 for w into the formula and simplify: $P = 2(3) + 2(2) = 6 + 4 = 10$. The correct answer is 10 feet. Part B asks you to find the volume of the rectangular box. The formula for the volume of a rectangular box, or prism, is $V = lwh$. Substitute 3 for l, 2 for w, and 2 for h, and simplify: $V = 3(2)(2) = 12$. The correct answer is 12 cubic feet.

INDEPENDENT PRACTICE

Read the problem. Write your answers.

7 Rajiv put $1,000 into a savings account that earns 5% interest each year.

   Part A: How much interest will he earn if he leaves the money in this account for 3 years?

   Answer: _____

   Part B: Rajiv invests $200 in another savings account that earns 4% interest. How many years will it take this savings account to earn $40 in interest?

   Answer: _____

UNIT 4 Algebra

INDEPENDENT PRACTICE

Read the problem. Write your answers.

8 The record low temperature in Islip, New York, during July was 50°F.

Part A: What is this record low temperature in degrees Celsius?

Show your work.

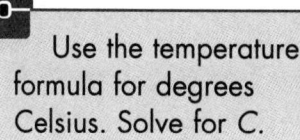

Use the temperature formula for degrees Celsius. Solve for C.

Answer: _____

Part B: The record high temperature in Islip, New York, during July was close to 40°C. What is this temperature in degrees Fahrenheit?

Show your work.

Answer: _____

UNIT 4 Algebra   95

Algebra Review

Read each problem. Circle the letter of the best answer.

1 Which expression shows *q* less than 5?

A 5 − q
B q − 5
C 5 ÷ q
D q ÷ 5

2 A picture frame is 24 inches long and 20 inches wide. What is its perimeter?

A 44 in.
B 88 in.
C 240 in.
D 480 in.

3 Pete wants to solve this equation for *a*.

$$4a = 72$$

What should Pete do to find the value of *a*?

A add 4 to 72
B subtract 4 from 72
C multiply 4 and 72
D divide 4 into 72

4 Deshana bought loaves of bread for $2 each and a jar of peanut butter for $3. Which expression can be used to find the total cost in dollars for the peanut butter and *b* loaves of bread?

A $2b + 3$
B $3b + 2$
C $2(b + 3)$
D $3(b + 2)$

5 Look at this expression.

$$5 \times (n − 2)^3$$

What is the value of this expression when *n* = 4?

A 12
B 14
C 30
D 40

6 A circular rug has a radius of 2 feet. What is the circumference of the rug? Use 3.14 for π.

A 5.14 feet
B 6.28 feet
C 12.56 feet
D 25.12 feet

7 Reeve gets 5 cents for each aluminum can he recycles. The equation below can be used to find the number of aluminum cans, *c*, he recycled one week.

$$5 \times c = 300$$

How many aluminum cans did Reeve recycle that week?

A 60
B 150
C 600
D 1,500

8 Leonard has $600 in a savings account with a 3% interest rate. How much interest will this savings account earn in 10 years?

A $1.80
B $18
C $180
D $1,800

Algebra Review

Read each problem. Write your answers.

9 A clothing store is holding a sale as shown on the sign at the right. Cindy buys 3 sweaters. Each sweater originally costs *d* dollars.

Part A: Write an expression that can be used to find Cindy's total cost for the 3 sweaters.

Answer: _____

Part B: What is Cindy's total cost, in dollars, if each sweater originally costs $50?

Answer: _____

10 The equation below can be used to find the number of tokens, *t*, Ethan had before playing a pinball game.

$$t + 8 = 43$$

Part A: What should Ethan do to find the value of *t*?

Answer: _____

Part B: What is the value of *t*?

Answer: _____

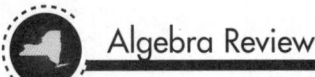

Algebra Review

Read the problem. Write your answers.

11 Clara rode her bike at an average speed of 6 miles per hour for 2 hours.

Part A: What total distance, in miles, did Clara ride her bike?

Answer: _____

Part B: In $1\frac{1}{2}$ hours, Stacy rode her bike the same distance as Clara. What was Stacy's average rate of speed, in miles per hour?

Show your work.

Answer: _____

Geometry

Geometry focuses on understanding properties of two- and three-dimensional figures. Sometimes plane figures are similar to each other, meaning they are the same shape but different sizes. You can find missing side lengths of similar figures using proportions. As you work with different figures, you will use many formulas to find measures, such as area and volume. Circles have measures, such as radius, diameter, and central angles, which are used in formulas. It is important to understand these formulas and to be able to use them correctly in different situations. When you plot points on a plane, you are using coordinate geometry.

This unit will help you answer test questions about geometry. There are seven lessons in this unit:

1. **Similar Triangles**   This lesson reviews how to find an unknown side length in a pair of similar triangles. You will also review how to find corresponding sides and angles in similar triangles.

2. **Area**   This lesson reviews formulas for finding area of rectangles, squares, parallelograms, triangles, and trapezoids. You will also use these formulas to find the areas of composite figures.

3. **Volume**   In this lesson, you will review using the formula for finding the volume of a rectangular prism.

4. **Circles**   This lesson reviews parts of a circle, such as the radius, the diameter, and chords. You will also use the circumference formula.

5. **Area of Circles**   In this lesson, you will review using the formula for finding the area of a circle.

6. **Central Angles and Sectors**   This lesson reviews central angles and sectors in circles. You will also review how to find the area of a sector by using the formula for the area of a circle.

7. **Coordinate Geometry**   In this lesson, you will review plotting points on a coordinate plane. You will also use a plane to find the perimeter of geometric figures.

Lesson 1: Similar Triangles

Indicator 6.G.1

similar triangles   corresponding angles   congruent   corresponding sides
congruent triangles   ratio   proportion

Similar triangles have the same shape but may not be the same size. **Corresponding angles** of similar triangles are **congruent**. They have the same measure. **Corresponding sides** are in proportion.

Triangle ABC is similar to triangle XYZ.

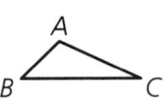

$\angle A \cong \angle X$ $\overline{AB} \sim \overline{XY}$

$\angle B \cong \angle Y$ $\overline{AC} \sim \overline{XZ}$

$\angle C \cong \angle Z$ $\overline{BC} \sim \overline{YZ}$

$$\frac{AB}{XY} = \frac{AC}{XZ} = \frac{BC}{YZ}$$

> **Congruent triangles** are the same size and shape. All congruent triangles are similar because they have the same shape.
>
> Similar triangles are **not** all congruent because they may be different sizes.

> The symbol $\cong$ means "is congruent to."
>
> The symbol $\sim$ means "is similar to."

A proportion can be set up to find the length of an unknown side in a pair of similar triangles.

These two triangles are similar. What is the length of the side marked x?

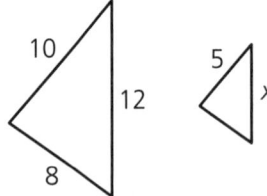

> A **ratio** is a comparison of two numbers.
>
> A **proportion** shows that two ratios are equal. These ratios are equivalent fractions.

1. Set up a proportion using the corresponding sides: $\frac{10}{5} = \frac{12}{x}$.

2. Cross multiply: $10x = 5 \times 12 \rightarrow 10x = 60$.

3. Divide each side by 10: $10x \div 10 = 60 \div 10 \rightarrow x = 6$.
   The side marked x is 6 units long.

GUIDED PRACTICE

Try this sample multiple-choice problem.

S These two triangles are similar.

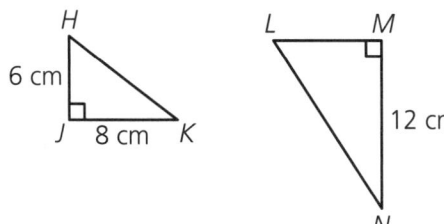

What is the length of $\overline{LM}$?

- **A** 8 cm
- **B** 9 cm
- **C** 10 cm
- **D** 16 cm

> This problem asks you to find the length of $\overline{LM}$. The shortest sides, $\overline{HJ}$ and $\overline{LM}$, correspond. The middle length sides, $\overline{JK}$ and $\overline{MN}$, also correspond. Use their measures to set up a proportion: $\frac{6}{LM} = \frac{8}{12}$. Cross multiply and solve: $8 \times LM = 6 \times 12 \rightarrow 8 \times LM = 72$, so $LM = 9$. The correct answer is B.

INDEPENDENT PRACTICE

Read each problem. Circle the letter of the best answer.

Use these similar triangles to answer questions 1 and 2.

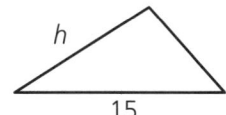 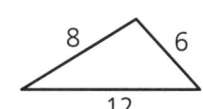

1 Which proportion can be used to find the length of the side labeled *h*?

- **A** $\frac{h}{6} = \frac{12}{15}$
- **B** $\frac{h}{6} = \frac{15}{12}$
- **C** $\frac{h}{8} = \frac{12}{15}$
- **D** $\frac{h}{8} = \frac{15}{12}$

2 What is the length, in units, of the side labeled *h*?

- **A** 6.4
- **B** 7.5
- **C** 10
- **D** 12

3 Triangle *NOP* is similar to triangle *QRS*.

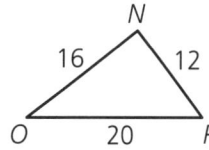 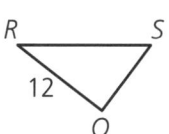

Which equation can be used to find the length of $\overline{RS}$?

- **A** $12(RS) = 16 \times 12$
- **B** $12(RS) = 20 \times 12$
- **C** $16(RS) = 12 \times 12$
- **D** $16(RS) = 20 \times 12$

4 Triangles *DEG* and *EFG* are similar.

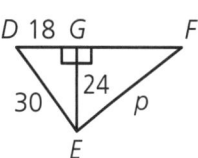

What is the value of *p*?

- **A** 36
- **B** 40
- **C** 48
- **D** 60

UNIT 5 Geometry

GUIDED PRACTICE

Try this sample constructed response problem.

S Look at these similar triangles.

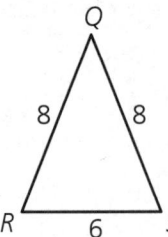 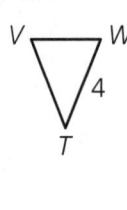

Part A: Write a proportion that can be used to find the length of $\overline{VW}$.

Answer: $\dfrac{8}{4} = \dfrac{6}{VW}$

Part B: What is the length of $\overline{VW}$?

Answer: 3 units

Part A asks you to write a proportion to find the length of $\overline{VW}$. First identify the corresponding sides. The two longer sides of triangle QRS are both 8 units long. These sides correspond to the longer sides of triangle TVW, $\overline{TV}$ and $\overline{TW}$. The shortest sides of each triangle, $\overline{RS}$ and $\overline{VW}$ must also correspond. Set up a proportion using the corresponding sides. The correct answer is $\dfrac{8}{4} = \dfrac{6}{VW}$. Part B asks you to find the length of $\overline{VW}$. Cross multiply and solve for VW: $8 \times VW = 24$, so $VW = 3$. The correct answer is 3 units.

INDEPENDENT PRACTICE

Read the problem. Write your answers.

5 These two triangles are similar.

What is the length of the side labeled *k*?

Show your work.

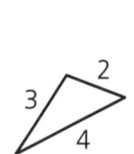 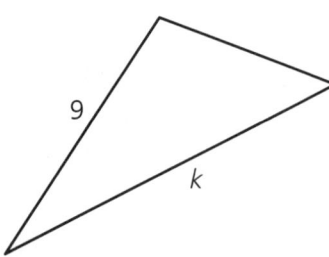

Answer: _____

UNIT 5 Geometry

Read the problem. Write your answers.

6 Triangles *CDF* and *GHJ* are similar.

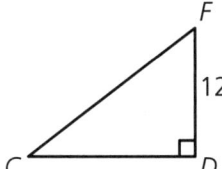 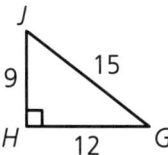

Part A: Write a proportion that can be used to find the length of $\overline{CD}$.

Find the sides that correspond to each other.

Answer: _____

What is the length, in units, of $\overline{CD}$?

Answer: _____

Part B: What is the perimeter, in units, of triangle *CDF*?

Answer: _____

Area

Indicators 6.G.2, 3

area   formula   rectangle   parallelogram   triangle
trapezoid   square   composite figure

Area is a measure of the number of square units inside a figure. A **formula** can be used to find the area of many figures.

Area of a **rectangle**
lw = length × width

Area of a **parallelogram**
bh = base × height

A **square** is a rectangle with 4 equal sides. The area of a square is s^2 = side × side.

Area of a **triangle**
$\frac{1}{2}bh = \frac{1}{2} \times$ base × height

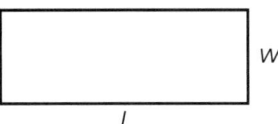

Area of a **trapezoid**
$\frac{1}{2} \times (b_1 + b_2) \times h =$
$\frac{1}{2} \times (base_1 + base_2) \times height$

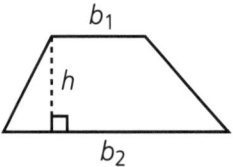

The bases of a trapezoid are the sides that are parallel.

To find the area of a figure, substitute the given values into the appropriate formula.

What is the area of this triangle?

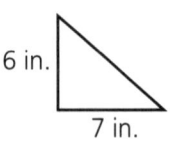
6 in.
7 in.

1. Determine the correct area formula:
   Area of a triangle = $\frac{1}{2}bh$.

2. Substitute the given values for b and h. Since b = 7 in. and h = 6 in., area = $\frac{1}{2} \times 7 \times 6$.

3. Simplify: $\frac{1}{2} \times 7 \times 6 = 21$ in.^{2}.

A **composite figure** is made up of more than one figure.

The area of a composite figure is the sum of the areas of each polygon that make up the figure.

UNIT 5 Geometry

GUIDED PRACTICE

Try this sample multiple-choice problem.

S The diagram shows a bathroom floor.

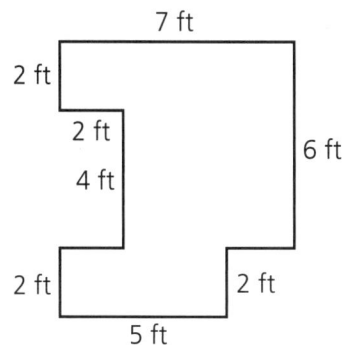

What is the area of the floor?

A 30 sq ft   C 42 sq ft

B 35 sq ft   D 44 sq ft

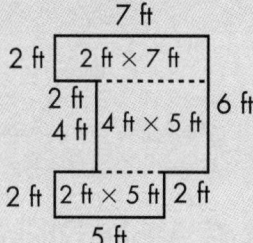

This problem asks you to find the area of an irregular, or composite, figure. First divide the figure into rectangles.

Find the area of each rectangle by multiplying length × width: 2 × 7 = 14, 4 × 5 = 20, and 2 × 5 = 10. Then add the areas of the rectangles to find the total area: 14 + 20 + 10 = 44. The correct answer is D.

INDEPENDENT PRACTICE

Read each problem. Circle the letter of the best answer.

1 Each side of a square playground has a length of 40 yards. What is the area of the playground?

A 160 yd²   C 1,600 yd²

B 400 yd²   D 4,000 yd²

2 What is the area of the figure below?

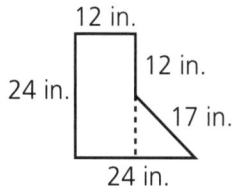

A 288 sq in.   C 390 sq in.

B 360 sq in.   D 432 sq in.

3 What is the area of this trapezoid?

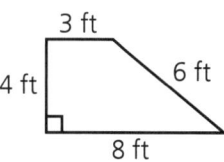

A 22 ft²   C 32 ft²

B 24 ft²   D 44 ft²

4 What is the area of this triangle?

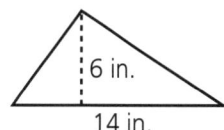

A 20 in.²   C 56 in.²

B 42 in.²   D 84 in.²

GUIDED PRACTICE

Try this sample constructed response problem.

S Each side of the hexagon shown below measures 8 feet.

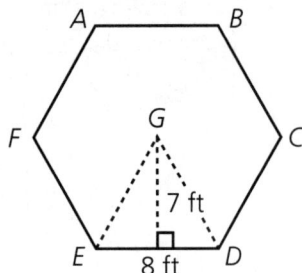

Part A: What is the area, in square feet, of triangle *GED*?

Answer: _____28 square feet_____

Part B: What is the area, in square feet, of hexagon *ABCDEF*?

Answer: _____168 square feet_____

> Part A asks you to find the area of a triangle with a base of 8 feet and a height of 7 feet. Substitute these values into the area formula for a triangle, $A = \frac{1}{2}bh$: $\frac{1}{2} \times 8 \times 7 = 28$. The correct answer is 28 square feet. Part B asks you to find the area of the entire hexagon. Since each side of the hexagon is the same, the hexagon can be divided into 6 triangles that have the same area as triangle *GED*. So, the area of the hexagon is $6 \times 28 = 168$. The correct answer is 168 square feet.

INDEPENDENT PRACTICE

Read the problem. Write your answers.

5 Look at this parallelogram.

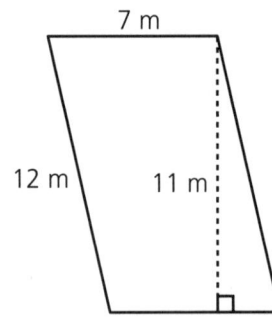

Part A: What is the height of this parallelogram?

Answer: _____

Part B: What is the area, in square meters, of this parallelogram?

Answer: _____

UNIT 5 Geometry

INDEPENDENT PRACTICE

Read the problem. Write your answers.

6 The figure below is made up of two parallelograms.

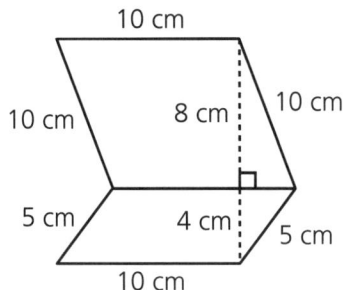

Part A: What formula should be used to find the area of each parallelogram?

Answer: _____

Part B: What is the area, in square centimeters, of the entire figure?

Show your work.

The dashed lines show the height of each parallelogram.

Answer: _____

UNIT 5 Geometry

LESSON 3

Volume

Indicator 6.G.4

volume   cubic units   rectangular prism   cube   prism

The **volume** of any solid figure is a measure of how much space is inside. Volume is found by counting **cubic units.**

Bert filled this box with cubes of the same size.

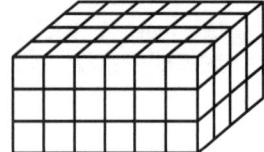

> Volume is *always* expressed in cubic units.
> Cubic feet = ft³
> Cubic meters = m³

What is the volume of this rectangular prism?

1. Count the number of cubes in the one layer: 24.

2. Multiply by the number of layers: 3 layers × 24 cubes per layer = 72. The volume of this prism is 72 cubic units.

To find the volume of a **rectangular prism** without counting the number of cubic units, use the volume formula. The volume of a rectangular prism is $V = length \times width \times height$ or $V = lwh$.

What is the volume of this rectangular prism?

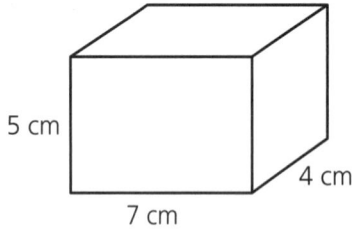

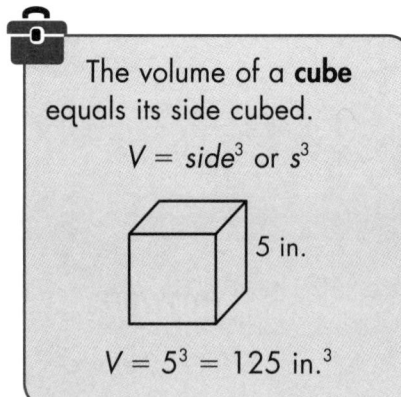

> The volume of a **cube** equals its side cubed.
> $V = side^3$ or s^3
>
> $V = 5^3 = 125$ in.³

1. Substitute the measures for each dimension into the volume formula for a prism: $V = 7 \times 4 \times 5$.

2. Multiply: $V = 7 \times 4 \times 5 = 140$. The volume of the rectangular prism is 140 cm³.

> The volume of any **prism** equals the product of the area of its base and its height.
> $V = (area\ of\ base) \times height$

UNIT 5 Geometry

© The Continental Press, Inc.   Do not duplicate.

GUIDED PRACTICE

Try this sample multiple-choice problem.

S Shayna wants to use this shoebox to store school supplies.

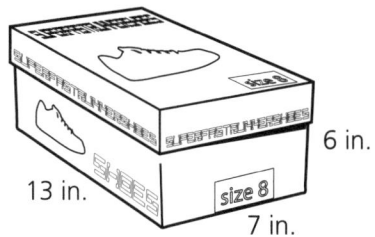

How much space is inside the shoebox?

A 26 in.³   C 169 in.³

B 52 in.³   D 546 in.³

> This problem asks you to find the volume of a rectangular prism with a length of 13 inches, a width of 7 inches, and a height of 6 inches. To do this, substitute these values into the volume formula $V = lwh$ and multiply: $V = 13 \times 7 \times 6 = 546$. The correct answer is D.

INDEPENDENT PRACTICE

Read each problem. Circle the letter of the best answer.

1 A cube measures 2 meters on each side. What is the volume of the cube?

A 6 m³

B 8 m³

C 12 m³

D 16 m³

2 How many cubic units fit into this rectangular prism?

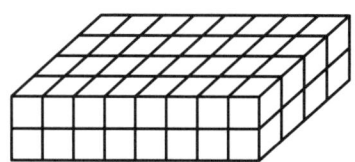

A 24

B 32

C 48

D 64

3 Martina started filling this container with 1-inch blocks.

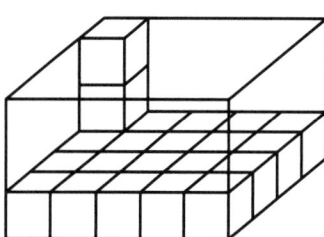

What is the total volume of this container when it is filled with these blocks?

A 16 in.³   C 60 in.³

B 32 in.³   D 80 in.³

4 The area of the floor in a storage closet is 30 square feet. The closet is 8 feet tall. What is the total amount of space inside the closet?

A 240 ft³   C 1,920 ft³

B 320 ft³   D 7,200 ft³

UNIT 5 Geometry

GUIDED PRACTICE

Try this sample constructed response problem.

S A cubic yard is 3 feet long by 3 feet wide by 3 feet tall.

Part A: How many cubic feet are in 1 cubic yard?

Answer: _____27 cubic feet_____

Part B: Landscaping mulch costs $1 per cubic foot. What is the cost for 3 **cubic yards** of landscaping mulch?

Answer: _____$81_____

> Part A asks you to find the number of cubic feet in 1 cubic yard. To do this, multiply 3 feet by 3 feet by 3 feet: $3 \times 3 \times 3 = 27$. The correct answer is 27 cubic feet. Part B asks you to find the cost of 3 cubic yards of mulch. You know that one cubic foot costs $1. Since there are 27 cubic feet in 1 cubic yards, 1 cubic yard costs $27. So, 3 cubic yards costs 3 times as much: $3 \times 27 = 81$. The correct answer is $81.

INDEPENDENT PRACTICE

Read the problem. Write your answers.

5 A garbage can is in the shape of a rectangular prism. The area of the base of the garbage can is 480 square inches. The height of the garbage can is 30 inches.

How many cubic inches of space are inside this garbage can?

Answer: _____

Explain how you know your answer is correct.

_____
_____
_____

UNIT 5 Geometry

© The Continental Press, Inc. Do not duplicate.

INDEPENDENT PRACTICE

Read the problem. Write your answers.

6 Look at the box below.

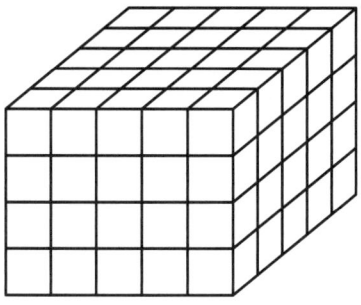

Part A: What is the volume, in cubic units, of this box? Show your work.

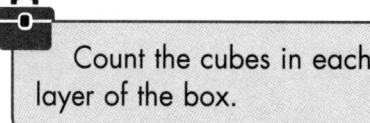
Count the cubes in each layer of the box.

Answer: _____

Part B: A second box with different dimensions has the same volume as the box shown above. What could be the dimensions of the second box?

Answer: _____

Explain how you know your answer is correct.

_____
_____
_____
_____

Circles

Indicators 6.G.5, 6, 7, 9

circle   radius(r)   diameter (d)   chord   circumference (C)

A **circle** is the set of all points that are the same distance from a center point.

The **radius** is the distance from the center of the circle to any point on the circle.

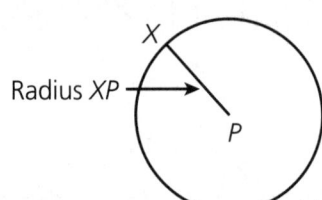

The plural of *radius* is *radii*. All radii in a given circle are the same length.

The **diameter** is the distance from one end of a circle to the other, through the center of the circle. The diameter is **always** twice the length of the radius in a circle.

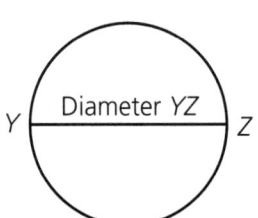

The radius of a circle is **always** half the length of the diameter.

$r = \frac{1}{2}d$

A **chord** is a line segment that joins **any** two points on a circle. The diameter is the longest chord in a circle.

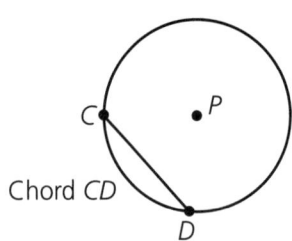

The number π (pi) is the ratio of the circumference of a circle to its diameter. Its value is approximately 3.14.

The **circumference (C)** is the distance around a circle. The formula for the circumference of a circle is $C = 2\pi r$ or $C = \pi d$.

  The diameter of a circle is 15 inches. What is the circumference of the circle?

  1. Write the circumference formula: $C = \pi d$.

  2. Substitute 3.14 for π and the value of the diameter for *d*:
     $C = 3.14 \times 15$.

  3. Multiply: $3.14 \times 15 = 47.1$. The circumference of the circle is 47.1 inches.

112   UNIT 5   Geometry

© The Continental Press, Inc.   Do not duplicate.

GUIDED PRACTICE

Try this sample multiple-choice problem.

S Which equation can be used to find the circumference of this circle?

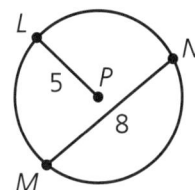

- **A** $C = 3.14 \times 5$
- **B** $C = 3.14 \times 8$
- **C** $C = 2 \times 3.14 \times 5$
- **D** $C = 2 \times 3.14 \times 8$

This problem asks you to identify the correct circumference equation. The circle has a radius of 5 units and a chord that is 8 units. The chord is **not** the diameter since it does not go through the center of the circle. The circumference formula that uses the radius is $C = 2\pi r$ or $C = 2 \times 3.14 \times r$. Substitute 5 for the radius. The correct answer is C.

INDEPENDENT PRACTICE

Read each problem. Circle the letter of the best answer.

1 The diameter of a juice glass is 50 mm. What is the radius of the glass?

- **A** 10 mm
- **B** 25 mm
- **C** 50 mm
- **D** 100 mm

2 Which name **best** describes $\overline{JL}$ in this circle?

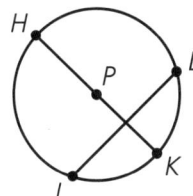

- **A** chord
- **B** radius
- **C** diameter
- **D** circumference

3 A circular baking pan has a diameter of 10 inches. What is the circumference of the baking pan?

- **A** 15.7 inches
- **B** 31.4 inches
- **C** 50.0 inches
- **D** 62.8 inches

4 Which statement **best** describes $\overline{SR}$ in the circle below?

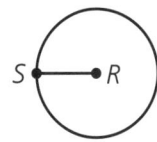

- **A** It is a chord only.
- **B** It is a radius only.
- **C** It is both a chord and a diameter.
- **D** It is both a radius and a diameter.

5 Which statement is true of any circle?

- **A** Every chord is also a diameter.
- **B** The radius is twice as long as the diameter.
- **C** The product of the radius and 3.14 is the circumference.
- **D** The value of the circumference divided by the diameter is approximately 3.14.

UNIT 5 Geometry

GUIDED PRACTICE

Try this sample constructed response problem.

S The radius of a circular movie reel is 6 centimeters.

Part A: What is the length of the diameter of this movie reel?

Answer: _____12 centimeters_____

Part B: What is the distance around this movie reel?

Answer: _____37.68 centimeters_____

> Part A asks you to find the diameter of a circle given the radius. The diameter is twice as long as the radius. To find the length of the diameter, multiply the radius by 2: 2 × 6 = 12. The correct answer is 12 centimeters. Part B asks you to find the circumference of the circle. The circumference formula is C = 2πr. Substitute the radius length for r and 3.14 for π into the formula and multiply: C = 2 × 3.14 × 6 = 37.68. The correct answer is 37.68 centimeters.

INDEPENDENT PRACTICE

Read the problem. Write your answers.

6 Maureen drew this circle.

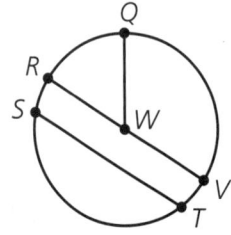

Part A: What name describes $\overline{ST}$?

Answer: _____

Part B: List all the given line segments that are radii of this circle.

Answer: _____

114   **UNIT 5** Geometry

INDEPENDENT PRACTICE

Read the problem. Write your answers.

7 A circular dining room table has a radius of 4 feet.

Part A: What is the distance, in feet, around the edge of this table?

Show your work.

> The circumference is the distance around a circular object.

Answer: _____

Part B: What is the diameter, in feet, of this dining room table?

Answer: _____

Explain how you know your answer is correct.

_____
_____
_____
_____

Area of Circles

Indicator 6.G.7

area (A)   pi (π)   radius   diameter

The **area** of a circle is the number of square units inside it. The formula for the area of a circle is $A = \pi r^2$.

The approximate value of the number π **(pi)** is 3.14.

A horse can roam in a circular field that has a radius of 6 meters.

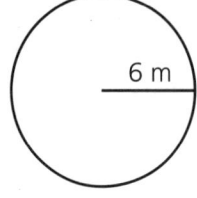

What is the area of this field?

1. Substitute the radius into the area formula: $A = \pi r^2 = 3.14 \times (6)^2$.

2. Simplify: $3.14 \times (6)^2 = 3.14 \times 36 = 113.04$. The area of the circular field is 113.04 m^2.

Area is *always* measured in square units.
Square feet = ft^2
Square millimeters = mm^2

Sometimes you will know the diameter of a circle and need to find the area.

What is the area of this circle?

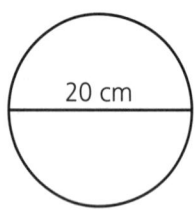

The **radius** of a circle is half the diameter.
$r = \frac{1}{2}d$

The **diameter** of a circle is twice the radius.
$d = 2r$

1. Find the radius: $r = \frac{1}{2}d$, so $r = \frac{1}{2} \times 20 = 10$.

2. Substitute the radius into the area formula: $A = \pi r^2 = 3.14 \times (10)^2$.

3. Simplify: $3.14 \times (10)^2 = 3.14 \times 100 = 314$. The area of the circle is 314 cm^2.

UNIT 5 Geometry

GUIDED PRACTICE

Try this sample multiple-choice problem.

S The circular lid of a jar has a diameter of 3 inches. What is the area of the lid?

A 7.065 in.²

B 9.42 in.²

C 18.84 in.²

D 28.26 in.²

> This problem asks you to find the area of a circle with a diameter of 3 inches. First find the length of the radius: $r = \frac{1}{2}d = \frac{1}{2} \times 3 = 1.5$. Then substitute the values for π and the radius into the area formula and simplify: $A = \pi r^2 = 3.14 \times (1.5)^2 = 3.14 \times 2.25 = 7.065$. The correct answer is A.

INDEPENDENT PRACTICE

Read each problem. Circle the letter of the best answer.

1 A circle has a radius of 4 meters. Which expression can be used to find the area of this circle?

A 3.14×2^2

B 3.14×4^2

C $2 \times 3.14 \times 2$

D $2 \times 3.14 \times 4$

2 The wheel of an inline skate has a radius of 40 millimeters. What is its area?

A 125.6 mm²

B 251.2 mm²

C 1,256 mm²

D 5,024 mm²

3 Which expression gives the correct area of this circle?

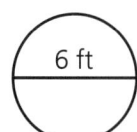

A 3.14×3^2

B 3.14×6^2

C 3.14×6

D 3.14×12

4 A circular clock face has a diameter of 10 inches. What is its area?

A 31.4 in.²

B 62.8 in.²

C 78.5 in.²

D 314 in.²

5 A pie plate measures 8 inches across the bottom. What is the area of the bottom of this pie plate?

A 25.12 in.²

B 50.24 in.²

C 100.48 in.²

D 200.96 in.²

6 The diameter of a circle is 12 units. Which expression can be used to find its area?

A 3.14×12

B $3.14 \times \left(\frac{12}{2}\right)$

C 3.14×12^2

D $3.14 \times \left(\frac{12}{2}\right)^2$

UNIT 5 Geometry

 GUIDED PRACTICE

Try this sample constructed response problem.

S Look at this circle.

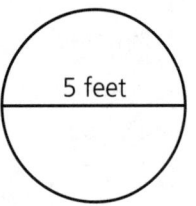

Part A: What is the radius, in feet, of this circle?

Answer: __2.5 feet__

Part B: What is the area, in square feet, of this circle?

Answer: __19.625 square feet__

> Part A asks you to find the radius of a circle with a diameter of 5 feet. The radius is half the diameter, or $\frac{1}{2} \times 5 = 2.5$. The correct answer is 2.5 feet. Part B asks you to find the area of the circle. Substitute the values for π and the radius into the area formula and simplify: $A = \pi r^2 = 3.14 \times (2.5)^2 = 3.14 \times 6.25 = 19.625$. The correct answer is 19.625 square feet.

INDEPENDENT PRACTICE

Read the problem. Write your answers.

7 A circular pond has a radius of 30 feet. What is the area, in square feet, of this pond?

Show your work.

Answer: _____

UNIT 5 Geometry

INDEPENDENT PRACTICE

Read the problem. Write your answers.

8 Circular pizzas at Renaldo's Pizza Shop come in these two sizes.

> Small Pizza—9-inch diameter
>
> Large Pizza—18-inch diameter

Part A: What is the area, in square inches, of the small pizza?

Show your work.

Answer: _____

Part B: Renaldo's prices each pizza by the square inch. Would you expect the price of the large pizza to be twice the price of a small pizza?

> Find the area of the large pizza. Compare it to the area of the small pizza.

Answer: _____

Explain how you know.

_____
_____
_____
_____

UNIT 5 Geometry

Lesson 6: Central Angles and Sectors

Indicators 6.G.5, 8

central angle   sector   area of a sector

A **central angle** of a circle is an angle whose vertex is at the center of the circle. A **sector** of a circle is the interior part of the circle defined by a central angle.

A circle has a total of 360°.

The center of this circle is point R. A central angle is ∠QRS. It measures 135°. The shaded part is sector QRS.

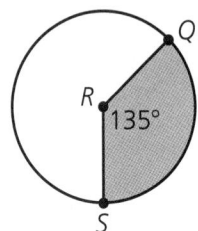

You can find the **area of a sector** of a circle by applying the formula for the area of a circle.

Sectors make up the sections of a circle graph.

Radius QR in circle R above is 5 centimeters long. What is the area of sector QRS?

1. Find the fraction of the circle represented by the sector. To do this, write a fraction with the measure of the central angle in the numerator and 360° in the denominator. Then reduce the fraction to lowest terms: $\frac{135}{360} = \frac{27}{72} = \frac{9}{24} = \frac{3}{8}$. Central angle QRS is $\frac{3}{8}$ of circle R.

2. Find the area of the entire circle: $A = \pi r^2 = 3.14 \times (5)^2 = 3.14 \times 25 = 78.5$ cm².

3. Multiply the area of the full circle by the fraction of the circle represented by the sector: $\frac{3}{8} \times 78.5 = 29.4375$. To the nearest tenth of a square centimeter, the area of sector QRS is 29.4 cm².

The formula for the area of a circle is $A = \pi r^2$.

The symbol π stands for the number pi. Its approximate value is 3.14.

UNIT 5 Geometry

GUIDED PRACTICE

Try this sample multiple-choice problem.

S Circle *T* has a diameter of 20 inches and a central angle that measures 120°. What is the area of the sector of circle *T* formed by this central angle?

- **A** 104.7 sq in.
- **B** 314 sq in.
- **C** 418.7 sq in.
- **D** 1,256 sq in.

> This problem asks you to find the area of a 120° sector. First find the fraction of the circle represented by the sector: $\frac{120}{360} = \frac{1}{3}$. Then find the area of the entire circle. Since the diameter is 20 inches, the radius is 10 inches: $A = 3.14 \times (10)^2 = 3.14 \times 100 = 314$ square inches. Finally, find the fraction of the entire area represented by the sector: $\frac{1}{3} \times 314 = 104.7$. The correct answer is A.

INDEPENDENT PRACTICE

Read each problem. Circle the letter of the best answer.

1 Which circle shows central angle *WXY*?

A 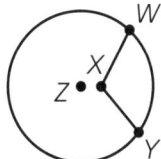 C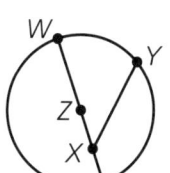

B 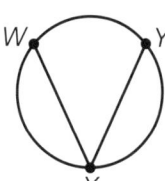 D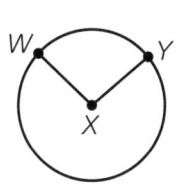

2 Central angle *NOP* in circle *O* measures 90°. The radius of this circle is 4 meters. What is the area of sector *NOP*?

- **A** 3.14 m²
- **B** 12.56 m²
- **C** 25.12 m²
- **D** 50.24 m²

3 Central angle *JKL* in circle *K* measures 150°. The radius measures 3 feet. What is the area of sector *JKL*, to the nearest square foot?

- **A** 12 ft²
- **B** 15 ft²
- **C** 19 ft²
- **D** 28 ft²

4 Which angle is **not** the name of a central angle in circle *D*?

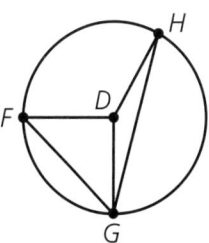

- **A** ∠FDG
- **B** ∠HDG
- **C** ∠FGH
- **D** ∠HDF

UNIT 5 Geometry

GUIDED PRACTICE

Try this sample constructed response problem.

S Circle A is shown below.

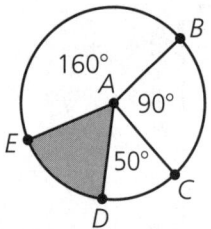

Part A: What is the name of the central angle surrounding the shaded section?

Answer: _____∠EAD_____

Part B: Circle A has a radius 6 yards long. What is the area, in square yards, of the shaded section?

Answer: _____18.84 sq yd_____

> Part A asks you to name a certain central angle. It has vertex A and endpoints E and D on the circle. The correct answer is ∠EAD. Part B asks you to find the area of the shaded section. First find the fraction of the circle that is shaded. Since a circle has a total of 360°, the degree measure of the shaded section is 360° − (160° + 90° + 50°) = 360° − 300° = 60°. This is the same as $\frac{60}{360} = \frac{1}{6}$ of the circle. Next find the area of the entire circle: $A = \pi r^2 = 3.14 \times (6)^2 = 3.14 \times 36 = 113.04$ sq yd. The area of the shaded section is $\frac{1}{6} \times 113.04 = 18.84$. The correct answer is 18.84 sq yd.

INDEPENDENT PRACTICE

Read the problem. Write your answers.

5 Point M is the center of this circle.

Name all the central angles in this circle.

Answer: _____

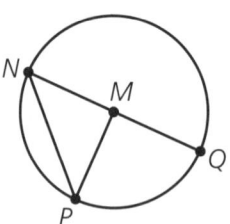

Explain how you know your answer is correct.

_____
_____
_____
_____

UNIT 5 Geometry

INDEPENDENT PRACTICE

Read the problem. Write your answers.

6 This dartboard is divided into 12 equal sections. The shaded parts represent the red sections of the dartboard.

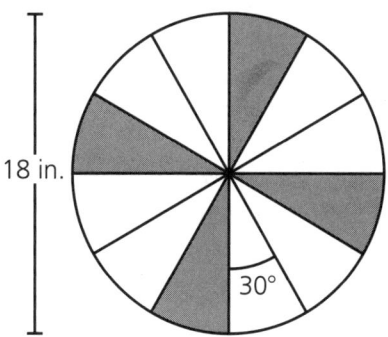

Part A: How many degrees of the dartboard make up of the red sections?

Answer: _____

Part B: What is the area, in square inches, of the red sections of the dartboard?

Show your work.

The radius of a circle is always half the diameter.

Answer: _____

UNIT 5 Geometry

Coordinate Geometry

Indicators 5.G.12, 13, 14

coordinate plane   ordered pair   x-axis   horizontal
y-axis   vertical   coordinates   origin

Points are plotted on a **coordinate plane.** The location of each point is named by an **ordered pair** of numbers. The first number in an ordered pair tells the distance the point is to the right of 0. The second number tells the distance the point is up from 0.

A coordinate plane has two axes. The **x-axis** is **horizontal.** The **y-axis** is **vertical.**

What is the location of point C on this coordinate plane?

1. Count the number of units the point is to the right of 0: 5.

2. Count the number of units the point is up from 0: 3. Point C is located at (5, 3) on the coordinate plane.

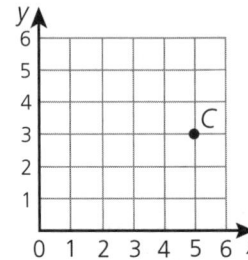

Ordered pairs are also called **coordinates.**

The point where the x-axis meets the y-axis is (0, 0). This is called the **origin.**

Sometimes you can connect points on a coordinate plane to form shapes.

Three points of rectangle QRST are plotted on the coordinate plane. What is the location of point T?

1. Identify the first number in the ordered pair. It is the same as the first number in the ordered pair for point S: 7.

2. Identify the second number in the ordered pair. It is the same as the second number in the ordered pair for point Q: 2. Point T in rectangle QRST is located at (7, 2).

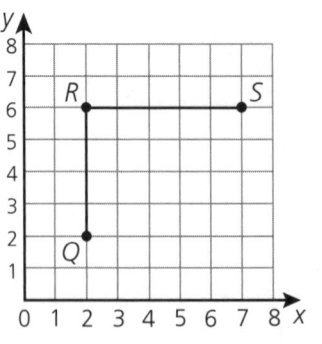

You can find the perimeter, or distance around, some shapes on a coordinate plane. The difference between the x-coordinates is the horizontal length. The difference between the y-coordinates is the vertical length.

UNIT 5 Geometry

© The Continental Press, Inc.   Do not duplicate.

GUIDED PRACTICE

Try this sample multiple-choice problem.

S   What is the perimeter of this square?

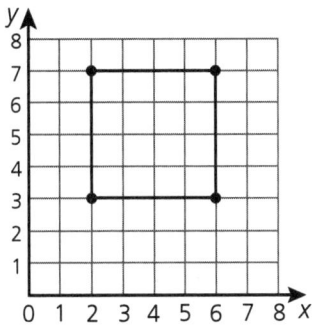

> This problem asks you to find the perimeter of the square. To do this, find a side length of the square by calculating the difference between the y-coordinates: $7 - 3 = 4$. Since each side length is the same, the perimeter of a square is $P = 4 \times side\ length = 4 \times 4 = 16$. The correct answer is D.

A   4 units   C   12 units

B   8 units   D   16 units

INDEPENDENT PRACTICE

Read each problem. Circle the letter of the best answer.

Use this coordinate plane to answer questions 1 and 2.

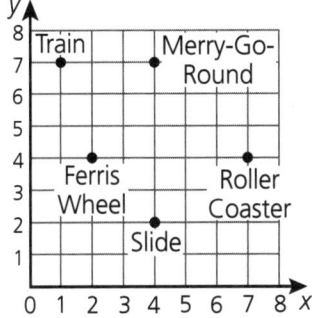

1 What are the coordinates where the train is located?

A   (1, 7)   C   (6, 2)

B   (2, 6)   D   (7, 1)

2 Which attraction has coordinates (7, 4)?

A   slide   C   roller coaster

B   Ferris wheel   D   merry-go-round

Use this coordinate plane to answer questions 3 and 4.

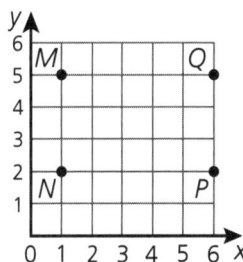

3 What figure is formed by drawing line segments to connect points M, N, and Q?

A   square   C   circle

B   triangle   D   rectangle

4 What is the perimeter of the figure formed by connecting points M, N, P, and Q?

A   8 units   C   16 units

B   12 units   D   24 units

UNIT 5 Geometry

GUIDED PRACTICE

Try this sample constructed response problem.

S Use this coordinate plane to answer the questions below.

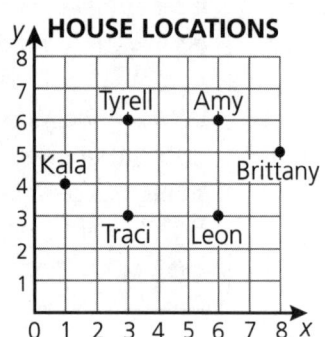

> Part A asks you to find the coordinates of Brittany's house. Brittany is 8 units to the right of 0 and 5 units up from 0. The correct answer is (8, 5). Part B asks you to identify the location with coordinates (3, 6). To do this, move 3 units to the right of 0. From there, move 6 units up. The correct answer is Tyrell.

Part A: What are the coordinates of Brittany's house?

Answer: ___(8, 5)___

Part B: Who has a house with coordinates (3, 6)?

Answer: ___Tyrell___

INDEPENDENT PRACTICE

Read the problem. Write your answers.

5 Use this coordinate plane to answer the questions below.

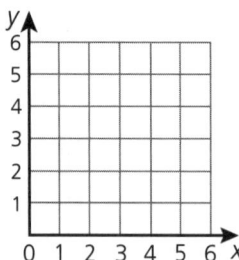

Part A: Plot and label these points on the coordinate plane: F (1, 5), G (5, 5), H (6, 2), J (2, 2). Then connect the points in order with line segments.

Part B: What type of figure did you make in part A?

Answer: _____

126 UNIT 5 Geometry

INDEPENDENT PRACTICE

Read the problem. Write your answers.

6 Use this coordinate plane to answer the questions below.

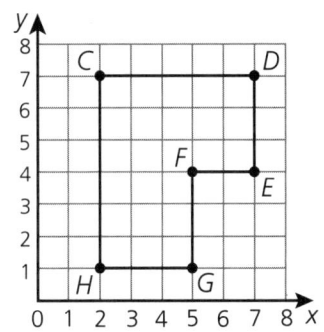

Part A: What are the coordinates of the endpoints of $\overline{CH}$?

Answer: _____

What is the length, in units, of $\overline{CH}$?

Answer: _____

Part B: What is the perimeter, in units, of the entire figure? Show your work.

> Find the lengths of each segment that make up the figure.

Answer: _____

Geometry Review

Read each problem. Circle the letter of the best answer.

1 A rectangular prism has a length of 8 inches, a width of 5 inches, and a height of 10 inches. What is the volume of this rectangular prism?

    A 23 in.^{3} **C** 200 in.^{3}

    B 46 in.^{3} **D** 400 in.^{3}

Use this circle to answer questions 2 and 3.

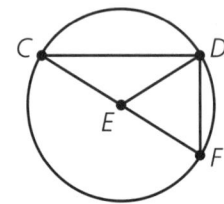

2 Which name describes $\overline{CD}$ in circle E?

    A chord   **C** diameter

    B radius   **D** circumference

3 Which of the following is a central angle in circle E?

    A $\angle CDF$ **C** $\angle DFE$

    B $\angle DEF$ **D** $\angle FCD$

4 The triangles below are similar.

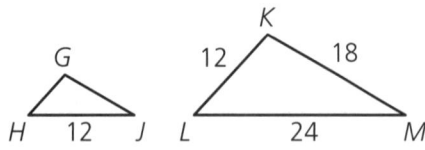

What is the length of $\overline{GJ}$?

    A 6 units   **C** 9 units

    B 8 units   **D** 10 units

Use this coordinate plane to help you answer question 5.

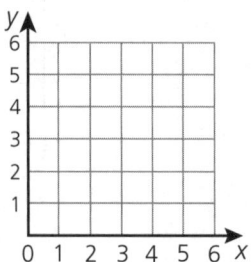

5 Figure $LMNP$ has coordinates L (1, 1), M (5, 1), N (4, 4), and P (1, 4). What type of figure is $LMNP$?

    A square   **C** rectangle

    B trapezoid   **D** parallelogram

6 Which expression represents the value of π in any circle?

    A radius × 2

    B area × radius^2

    C area ÷ diameter

    D circumference ÷ diameter

7 What is the area, to the nearest tenth square meter, of the shaded part of this circle?

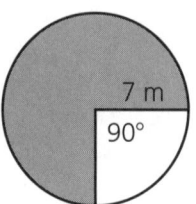

    A 33.0 m^2 **C** 115.4 m^2

    B 38.5 m^2 **D** 153.9 m^2

Geometry Review

Read each problem. Write your answers.

8 Greg plotted the three vertices of rectangle *QRST* shown.

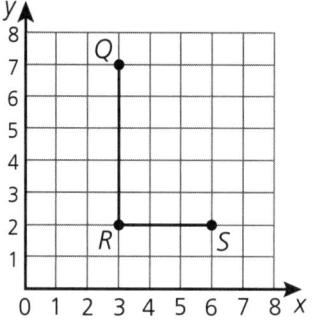

Part A: What are the coordinates of point *T*?

Answer: _____

Part B: What is the perimeter, in units, of rectangle *QRST*?

Answer: _____

9 Theresa drew this figure using triangles and rectangles.

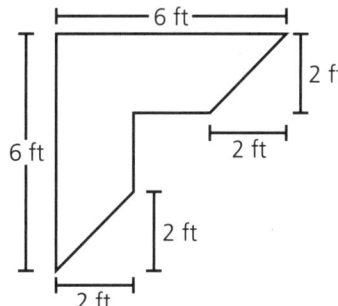

What is the area, in square feet, of this figure?

Show your work.

Answer: _____

UNIT 5 Geometry

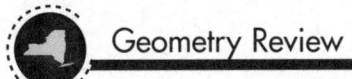

Geometry Review

Read the problem. Write your answers.

10 The diameter of this saucer is 12 centimeters long.

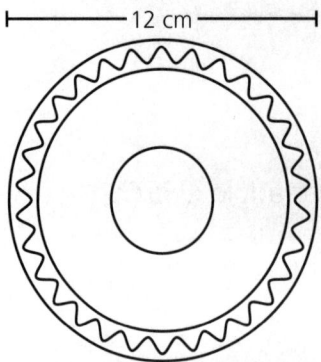

Part A: What is the length, in centimeters, of the radius of this saucer?

Answer: _____

Part B: What is the circumference, in centimeters, of the saucer?

Answer: _____

What is the area, in square centimeters, of the saucer?

Answer: _____

UNIT 5 Geometry

Unit 6: Measurement

Measurement is one of the most common uses of mathematics. The amount of juice in a pitcher and the amount of water in a swimming pool are examples of capacity. You can measure capacity with both customary and metric units. Sometimes you may need to change a smaller unit to a larger unit or a larger unit to a smaller unit. To do this, you must know equivalent units. Volume and capacity are related, so you can find the volume of a five-gallon bucket. Sometimes an exact measurement is not needed, but an estimate will do. Using the right technique, you can find an accurate estimate.

This unit will help you answer test questions about measurement. There are four lessons in this unit:

1. **Customary Units of Capacity** This lesson reviews customary units of capacity—cups, pints, quarts, and gallons. You will find equivalent units for customary measurements.

2. **Metric Units of Capacity** This lesson reviews metric units of capacity—milliliters and liters. You will find equivalent units for metric measurements.

3. **Relationship of Volume and Capacity** This lesson reviews how volume and capacity are related. You will find the volume and the capacity of figures using both customary and metric units.

4. **Estimating Measurements** In this lesson, you will review how to find an accurate estimate for volume and area of different figures, and for circumference of a circle.

Customary Units of Capacity

Indicators 6.M.2, 3, 6

capacity   cups (C)   pints (pt)   quarts (qt)   gallons (gal)

Customary units of **capacity** include **cups, pints, quarts,** and **gallons.**

2 cups = 1 pint   2 pints = 1 quart   4 quarts = 1 gallon

 Capacity is the measure of the amount of liquid a container holds.

To change larger units into smaller units, **multiply.**

How many quarts of orange juice are in a $\frac{1}{2}$-gallon container?

1. Identify the number of quarts in 1 gallon: 4 qt = 1 gal.

2. Multiply to change larger units into smaller units: $4 \times \frac{1}{2} =$ 2 quarts. There are 2 quarts in $\frac{1}{2}$ gallon.

To change smaller units into larger units, **divide.**

A recipe uses 3 cups of buttermilk. How many pints is this?

1. Identify the number of cups in 1 pint: 2 C = 1 pt.

2. Decide whether you should multiply or divide. Smaller units are being changed into larger units, so divide: $3 \div 2 = \frac{3}{2} = 1\frac{1}{2}$ pt. There are $1\frac{1}{2}$ pints in 3 cups of buttermilk.

Tools such as tablespoons, measuring cups, and gallon jugs are used to measure capacity.

UNIT 6 Measurement

© The Continental Press, Inc.   Do not duplicate.

GUIDED PRACTICE

Try this sample multiple-choice problem.

S Aldo drank 6 cups of water today. What fraction of a gallon is this?

A $\frac{1}{6}$ gal

B $\frac{1}{3}$ gal

C $\frac{3}{8}$ gal

D $\frac{3}{4}$ gal

> This problem asks you to find the gallon equivalent to 6 cups. First change cups to pints. There are 2 cups in 1 pint: 6 C ÷ 2 = 3 pt. Next, change pints to quarts. There are 2 pints in 1 quart: 3 pt ÷ 2 = $\frac{3}{2}$ qt. Finally, change quarts to gallons. There are 4 quarts in 1 gallon: $\frac{3}{2}$ ÷ 4 = $\frac{3}{2}$ × $\frac{1}{4}$ = $\frac{3}{8}$ gal. The correct answer is C.

INDEPENDENT PRACTICE

Read each problem. Circle the letter of the best answer.

1 Which tool would be **best** to measure the capacity of a container of syrup?

A scale

B teaspoon

C tape measure

D measuring cup

2 Which unit is **best** for measuring the capacity of a soda can?

A cups   C quarts

B pints   D gallons

3 A can of pineapple juice holds 3 quarts. How many gallons is this?

A $\frac{3}{4}$ C 6

B $1\frac{1}{2}$ D 12

4 Blaine is measuring capacity using gallons. Which of the following amounts could Blaine be measuring?

A broth added to a soup recipe

B milk added to a bowl of cereal

C water in a washing machine

D water in a bottle from a vending machine

5 Six quarts of soap will be poured into 1-pint containers. How many pint containers can be filled?

A 6   C 24

B 12   D 36

6 Which amount is the same as 4 gallons?

A 8 pints   C 16 pints

B 8 quarts   D 16 quarts

UNIT 6 Measurement

GUIDED PRACTICE

Try this sample constructed response problem.

S Deliah is washing her car.

Part A: Which unit is **best** for describing the amount of water she will use to wash her car?

Answer: ___gallons___

Part B: Deliah pours soap into a bucket that holds 6 quarts. How many cups does this bucket hold?

Answer: ___24___

> Part A asks you to identify the best unit for measuring the amount of water used to wash a car. This will be a relatively large amount, so a large unit would be best to use. The correct answer is gallons. Part B asks you to find the number of cups in 6 quarts. First change quarts to pints. There are 2 pints in 1 quart: 6 qt × 2 = 12 pt. Next, change pints to cups. There are 2 cups in 1 pint: 12 pt × 2 = 24 C. The correct answer is 24.

INDEPENDENT PRACTICE

Read the problem. Write your answers.

7 Look at these measures.

        15 cups   15 pints   1.5 gallons

List these measures in order from **smallest to largest**.

Show your work.

Answer: _____

UNIT 6 Measurement

INDEPENDENT PRACTICE

Read the problem. Write your answers.

8 Joey is making a glass of chocolate milk.

Part A: Which unit **best** describes the capacity of the glass?

Answer: _____

Part B: Joey has a gallon container of milk. He already used 2 pints. How many pints of milk are left?

Show your work.

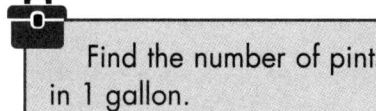

Find the number of pints in 1 gallon.

Answer: _____

Metric Units of Capacity

Indicators 6.M.4, 5, 6

capacity   milliliters (mL)   liters (L)

Metric units of **capacity** include **milliliters** and **liters.**

1,000 milliliters = 1 liter

To change larger units into smaller units, **multiply** or move the decimal point to the right. Put in zeros as needed.

How many milliliters are in 5 liters?

1. Identify the number of milliliters in 1 liter: 1,000 mL = 1 L.

2. Multiply to change larger units into smaller units: 5 × 1,000 = 5,000 mL. There are 5,000 milliliters in 5 liters.

Metric units are in multiples of 10. As a shortcut to multiplying or dividing by 10, you can move the decimal point.

To change smaller units into larger units, **divide** or move the decimal point to the left. Put in zeros as needed.

A bottle holds 600 milliliters of water. How many liters is this?

1. Identify the number of milliliters in 1 liter: 1,000 mL = 1 L.

2. Decide whether you should multiply or divide. Smaller units are being changed into larger units, so divide: 600 ÷ 1,000 = 0.600 = 0.6 L. There is 0.6 liter in 600 milliliters of water.

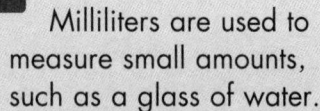

Milliliters are used to measure small amounts, such as a glass of water.

Liters are used to measure larger amounts, such as the water in a bathtub.

UNIT 6 Measurement

GUIDED PRACTICE

Try this sample multiple-choice problem.

S A chef buys $\frac{1}{2}$ liter of cream each week. How much cream does he buy in 4 weeks?

A   20 mL

B   200 mL

C   2,000 mL

D   20,000 mL

> This problem asks you to find the number of milliliters in 4 times $\frac{1}{2}$ liter. To do this, first change $\frac{1}{2}$ L to mL. There are 1,000 mL in 1 L: $\frac{1}{2}$ L × 1,000 = 500 mL. Then multiply this amount by 4 weeks: 4 weeks × 500 mL = 2,000 mL. The correct answer is C.

INDEPENDENT PRACTICE

Read each problem. Circle the letter of the best answer.

1 Which amount is **best** measured using liters?

A   liquid in an eyedropper

B   food coloring in a bottle

C   water a cat drinks in a day

D   juice sold at a snack bar in a day

2 A pitcher holds 1.2 liters of lemonade. How many milliliters does the pitcher hold?

A   0.00012

B   0.0012

C   1,200

D   12,000

3 How many liters are in a pot that holds 3,250 milliliters of soup?

A   0.325

B   3.25

C   32,500

D   3,250,000

4 How many liters of liquid are in this measuring container?

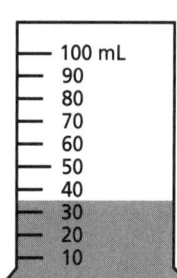

A   0.035

B   0.35

C   3,500

D   35,000

5 A water cooler contains 10 liters of water. Paper cups hold 200 milliliters. How many paper cups can be filled with the water in the cooler?

A   20

B   50

C   200

D   500

UNIT 6 Measurement

GUIDED PRACTICE

Try this sample constructed response problem.

S A 400-milliliter bottle of bleach costs $2. A 0.75-liter bottle of bleach costs $3.

Which size bottle is the better deal?

Answer: _____0.75-liter_____

Explain how you know your answer is correct.

> This problem asks you to find the bottle size that is the better deal. First change both bottles to the same units. There are 1,000 milliliters in 1 liter. To change liters to milliliters, multiply by 1,000: 0.75 L × 1,000 = 750 mL. Next, divide the costs by their corresponding bottle sizes. The one with the lower cost per milliliter is the better deal: $2 ÷ 400 mL = $0.005 per mL; $3 ÷ 750 mL = $0.004 per mL. Since $0.004 is smaller than $0.005, the better deal is the 0.75-liter bottle.

INDEPENDENT PRACTICE

Read the problem. Write your answers.

6 Ian has $\frac{1}{4}$ liter of a hot beverage. Manuel has 175 milliliters of the same hot beverage.

Part A: Who has more of the hot beverage—Ian or Manuel?

Answer: _____

Part B: How many milliliters more?

Answer: _____

INDEPENDENT PRACTICE

Read the problem. Write your answers.

7 Tricia has these two measuring containers of liquid.

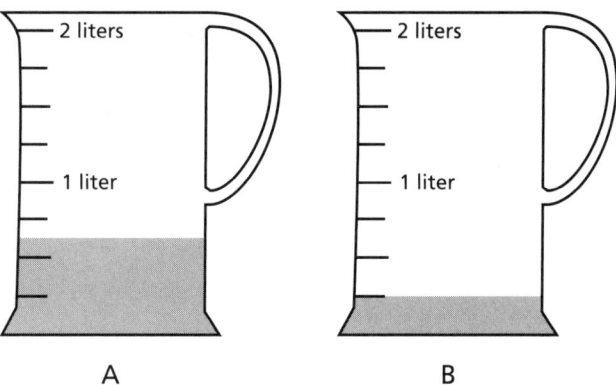

Part A: How many liters are in container A?

Answer: _____

Part B: Tricia pours the liquid from container A to container B. How many milliliters of liquid are now in container B?

Show your work.

> Be sure the denominators are the same before adding or subtracting fractions.

Answer: _____

LESSON 3: Relationship of Volume and Capacity

Indicator 6.M.1

volume   cubic inches   cubic centimeters

Capacity is a measure of **volume.** If you know the volume of a container, you can find the capacity of the container.

In the customary system of measurement, one gallon has a volume of 231 **cubic inches.**

About how many gallons of water can fit into this container?

1. Find the volume of the container in cubic inches. Use the volume formula for a rectangular prism, $V = lwh$:
$V = 10$ in. $\times$ 6 in. $\times$ 5 in. $= 300$ in.^{3}.

2. Divide by 231 to change cubic inches to gallons: $\frac{300}{231} = 1.2987... \approx 1.3$. About 1.3 gallons of water can fit into the container.

> Volume of a rectangular prism:
> $V = $ length $\times$ width $\times$ height $= lwh$

In the metric system of measurement, one liter has a volume of 1,000 **cubic centimeters.**

A rectangular tank is 200 centimeters long, 50 centimeters wide, and 50 centimeters tall. How many liters of water can this tank hold?

1. Find the volume of the tank in cubic centimeters. Use the volume formula for a rectangular prism, $V = lwh$: $V = 200$ cm $\times$ 50 cm $\times$ 50 cm $= 500{,}000$ cm^3.

2. Divide by 1,000 to change cubic centimeters to liters: $\frac{500{,}000}{1{,}000} = 500$ liters. The tank holds 500 liters of water.

> One cubic centimeter is the same as 1 milliliter. So, volume in cubic centimeters is the same as capacity in milliliters.
> 48 cm^3 = 48 mL

UNIT 6 Measurement

GUIDED PRACTICE

Try this sample multiple-choice problem.

S A rectangular can of oil is 12 cm long, 5 cm wide, and 15 cm tall. How many liters of oil fit into this can?

A 0.9

B 3.2

C 32

D 900

> This problem asks you to find the number of liters that fit into a rectangular-shaped can. To do this, first find the volume of the can: $V = 12 \text{ cm} \times 5 \text{ cm} \times 15 \text{ cm} = 900 \text{ cm}^3$. Then divide the volume by 1,000 to find the amount in liters: $900 \text{ cm} \div 1{,}000 = 0.9$ liter. The correct answer is A.

INDEPENDENT PRACTICE

Read each problem. Circle the letter of the best answer.

1 A container has a volume of 75 cubic centimeters. What is the capacity of the container in milliliters?

A 0.075 mL

B 7.5 mL

C 75 mL

D 75,000 mL

2 A water tank has a volume of 500,000 in.^{3}. About how many gallons of water can this tank hold?

A 0.00462   C 2,310

B 2,160   D 4,620

3 A carton of vegetable broth is 9 cm long, 6 cm wide, and 18 cm tall. What is the capacity, in milliliters, of this carton?

A 0.972 mL   C 972 mL

B 9.72 mL   D 972,000 mL

4 A box is 24 inches long, 18 inches wide, and 4 inches deep. What is the box's approximate capacity, in gallons?

A 7.5 gal   C 12.5 gal

B 9 gal   D 16 gal

5 A fish pond is in the shape of a rectangular prism. It is 60 inches long, 30 inches wide, and 20 inches deep. What is its approximate capacity, in gallons?

A 110 gal   C 235 gal

B 155 gal   D 360 gal

6 A storage chest is 120 cm long, 80 cm deep, and 50 cm tall. What is the capacity of the storage chest in liters?

A 480 L

B 4,800 L

C 48,000 L

D 480,000 L

UNIT 6 Measurement

GUIDED PRACTICE

Try this sample constructed response problem.

S The sink in Mr. Wallace's kitchen is shaped like a rectangular prism. It is 64 cm long, 50 cm wide, and 30 cm deep.

  Part A: What is the capacity of the sink in milliliters?

  Answer: _____96,000 mL_____

  Part B: What is the capacity of the sink in liters?

  Answer: ___96 L___

> Part A asks you to find the capacity of a sink in milliliters. Since 1 milliliter is the same as 1 cubic centimeter, the capacity in milliliters is the same as the volume in cubic centimeters. The volume of the sink is $V = lwh =$ 64 cm $\times$ 50 cm $\times$ 30 cm = 96,000 cm^3. The correct answer is 96,000 mL. Part B asks you to find the capacity in liters. Divide the volume by 1,000: 96,000 $\div$ 1,000 = 96. The correct answer is 96 L.

INDEPENDENT PRACTICE

Read the problem. Write your answers.

7 This container is filled halfway with water.

  Part A: What is the volume of this container in cubic centimeters?

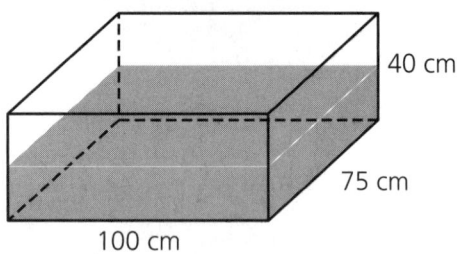

  Answer: _____

  Part B: How many liters of water are in the container?

  Answer: _____

INDEPENDENT PRACTICE

Read the problem. Write your answers.

8 The dimensions of a block of ice are shown below.

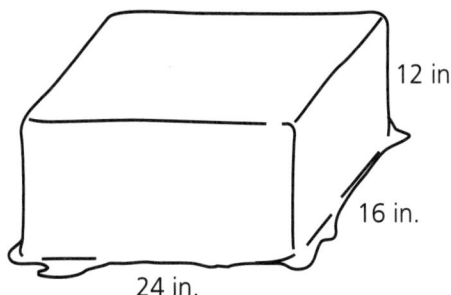

Part A: What is the volume, in cubic inches, of this ice block?

Show your work.

Answer: _____

Part B: What is the capacity, to the nearest tenth gallon, of water in this ice block?

Show your work.

> There are 231 cubic inches in one gallon.

Answer: _____

UNIT 6 Measurement

Lesson 4: Estimating Measurements

Indicators 6.M.7, 8, 9

estimate   round   area   volume   circumference   personal references

Some measurements do not need to be exact. You can use an **estimate.** One way to estimate is to **round** given measurements.

To estimate **area** of a rectangle, round the length and width.

What is the approximate area of a tabletop that is 5.25 feet long and 2.92 feet wide?

1. Round the length and width: 5.25 rounds down to 5 and 2.92 rounds up to 3.

2. Multiply the rounded length and width: 5 × 3 = 15. The tabletop has an area of approximately 15 square feet.

> To round a number, look at the digit to the right of the place you are rounding to. If the digit is 5 or greater, round *up.* If the digit is less than 5, round *down.*

To estimate **circumference** or area of a circle, round the radius or diameter.

What is the approximate circumference of a circle whose diameter is 8.83 meters?

1. Round the diameter: 8.83 rounds up to 9.

2. Round π: π rounds to 3.

3. Substitute the rounded values into the circumference formula: $C = \pi d = 3 \times 9 = 27$. The circumference is approximately 27 meters.

> The number π, which is approximately 3.14, can be rounded to 3 for a very rough estimate.

You can use **personal references** to decide whether or not estimated measurements are reasonable.

Vera estimates that there are 2 cups of soup in a can. Is her estimate reasonable?

Yes, a cup is about the size of a mug. The soup in a can would fit into 2 mugs. So, this estimate is reasonable.

> Some personal references for capacity:
> A mug is about 1 cup.
> A teaspoon is about 5 milliliters.
> A large soda bottle is about 2 liters.

UNIT 6 Measurement

© The Continental Press, Inc.   Do not duplicate.

GUIDED PRACTICE

Try this sample multiple-choice problem.

S A suitcase is 2.75 feet long, 0.83 foot tall, and 2.25 feet wide. What is the approximate volume of this suitcase?

   A   4 ft³

   B   6 ft³

   C   8 ft³

   D   12 ft³

> This problem asks you to estimate the volume of a suitcase. The volume of a rectangular prism equals length × width × height. To estimate the volume, first round each measurement: 2.75 rounds up to 3, 0.83 rounds up to 1, and 2.25 rounds down to 2. Then multiply the rounded values: 3 × 1 × 2 = 6. The correct answer is B.

INDEPENDENT PRACTICE

Read each problem. Circle the letter of the best answer.

1 Jim poured a capful of laundry detergent in with his clothes. What is the **most likely** capacity of a capful of detergent?

   A   1 cup   **C**   1 quart

   B   1 pint   **D**   1 gallon

2 Which is the **best** estimate for the area of this triangle? (Use $A = \frac{1}{2}bh$.)

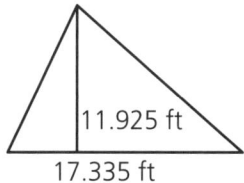

   A   75 sq ft   **C**   150 sq ft

   B   100 sq ft   **D**   200 sq ft

3 Which of these amounts **best** describes the amount of shampoo in a large bottle?

   A   40 mL   **C**   40 L

   B   400 mL   **D**   400 L

4 A circle has radius 43 mm. What is the **best** estimate for the area of the circle? (Use $A = \pi r^2$.)

   A   120 mm²

   B   240 mm²

   C   4,800 mm²

   D   7,500 mm²

5 The radius of a circle is 28.2 cm long. Rachel estimates that the circumference of this circle is about 2,700 cm. Is her estimate reasonable?

   A   Yes, it should be **about** 2,700 cm.

   B   No, it should be **about** 90 cm.

   C   No, it should be **about** 180 cm.

   D   No, it should be **about** 1,800 cm.

GUIDED PRACTICE

Try this sample constructed response problem.

S The length of a side of a square is 7.8 meters. Mandy estimates the area of the square to be about 48 square meters.

Is her estimate reasonable?

Answer: ___no___

Explain how you know your answer is correct.

> This problem asks you to determine if an estimate for the area of a square is reasonable. First round the length: 7.8 rounds up to 8. Then find the estimated area of the square. The formula for the area of a square is s^2, or $side^2$: $8^2 = 64$. A reasonable estimate for the area of this square would be 64 square meters. So, Mandy's estimate of 48 square meters is not reasonable.

INDEPENDENT PRACTICE

Read the problem. Write your answers.

6 A storage locker is 16.25 feet long, 10.33 feet wide, and 7.67 feet tall.

Part A: What are the rounded dimensions, in feet, of the storage locker?

Length: _____

Width: _____

Height: _____

Part B: What is a reasonable estimate for the volume, in cubic feet, of the storage locker?

Answer: _____

UNIT 6 Measurement

INDEPENDENT PRACTICE

Read the problem. Write your answers.

7 Giada is going swimming. She estimates that a swimming pool can hold up to 100 gallons of water.

   Part A: Is this estimate reasonable?

   Answer: _____

> Think of a personal reference that holds 1 gallon.

Explain how you know.

_____
_____
_____
_____

Part B: To rinse off after swimming, Giada takes a shower. What would be a reasonable estimate for the number of gallons of water Giada uses in the shower—2 gallons, 20 gallons, or 200 gallons?

   Answer: _____

UNIT 6 Measurement

Measurement Review

Read each problem. Circle the letter of the best answer.

1 Kristi wants to measure the amount of water that will be mixed with a packet of plant food. Which of these tools is **best** for her to use?

A

B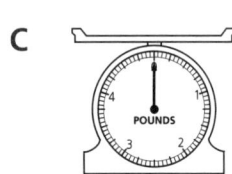

C

D

2 Which of the following is **best** measured using milliliters?

A length of a fingernail

B amount of water used to wash a dog

C heaviness of a bag of apples

D amount of food coloring for a recipe

3 A rectangular vase is 60 cm long, 25 cm wide, and 80 cm tall. What is its capacity, in milliliters?

A 120 mL

B 1,200 mL

C 120,000 mL

D 1,200,000 mL

4 Which is the **best** unit to measure the amount of water used to wash a load of dishes?

A cups C quarts

B pints D gallons

5 Which of the following is **most likely** to hold 200 milliliters when it is filled?

A a juice box C a bathtub

B a soda bottle D a garbage can

6 A circle has a diameter of 5.81 feet. What is the **best** estimate for the area of this circle? (Use $A = \pi r^2$.)

A 18 ft² C 75 ft²

B 27 ft² D 108 ft²

7 Vanessa mixed 800 milliliters of pineapple juice with 1,250 milliliters of orange juice to make a fruit punch. How many liters of fruit punch did Vanessa make?

A 2.05 C 20.5

B 2.50 D 25.0

8 Wendell filled his car's gas tank with 10.5 gallons of gas. How many pints of gas is that?

A 42 C 168

B 84 D 336

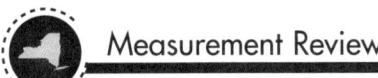

Measurement Review

Read each problem. Write your answers.

9 A triangular sign is shown at the right.

Mara estimates the area of the sign to be 85 square inches. Is this estimate reasonable?

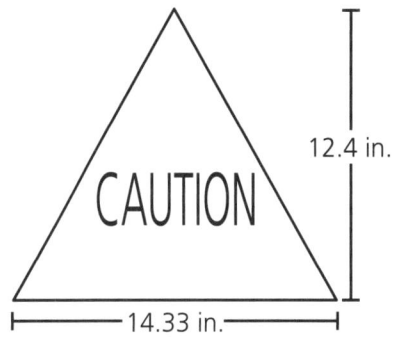

Answer: _____

Explain how you know.

_____
_____
_____
_____

10 A recipe uses 2 cups of milk.

Part A: How many pints of milk does this recipe use?

Answer: _____

Part B: Lakisha opens a one-gallon container of milk and pours out 2 cups for the recipe. How many quarts of milk does she have left?

Answer: _____

Measurement Review

Read the problem. Write your answers.

11 Look at this rectangular prism.

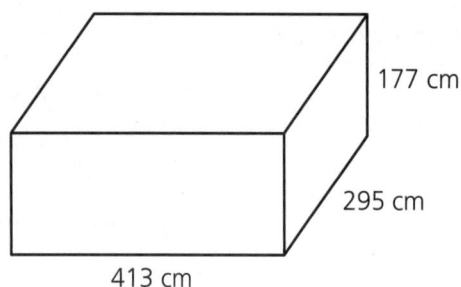

177 cm
295 cm
413 cm

Part A: Round the dimensions of this prism.

Length: _____

Width: _____

Height: _____

What is the approximate volume, in cubic centimeters, of the prism?

Answer: _____

Part B: What is the approximate capacity, in liters, of this prism?

Answer: _____

Statistics

Data is information. It can be organized in various kinds of tables and graphs to make it easier to understand and to use. You could use a bar graph to compare the number of e-mails you received from different people in a week. Your teacher might make a histogram to show the final grades for the students in your class. A newspaper might show how a stock value changed over time using a line graph. And you might see the survey results for an election displayed in a circle graph. As you look at data in tables and graphs, you can use different measures of central tendency and spread to better understand the data.

This unit will help you answer test questions about data and statistics. There are five lessons in this unit:

1. **Statistics**   This lesson reviews how to find the mean, median, mode, and range for given data sets.

2. **Bar Graphs**   This lesson reviews how to read and understand bar graphs. You will also use bar graphs to make predictions about data.

3. **Histograms**   In this lesson, you will review how to read and understand histograms. You will also draw conclusions and make predictions using histograms.

4. **Line Graphs**   In this lesson, you will review how to read and understand line graphs. You will use the trend in the graph to understand the data.

5. **Circle Graphs**   This lesson reviews how to read and understand circle graphs. You will find data that is missing from given circle graphs.

Lesson 1: Statistics

Indicators 6.S.5, 6

data statistics mean average median ordered data mode range

A **data** set is a collection of numerical values. Data values can be described using different **statistics**.

The **mean** of a data set is the average of all the values.

> Rowan's test grades are 92, 89, 95, 89, 90. What is the mean of these grades?
>
> 1. Add all the values: 92 + 89 + 95 + 89 + 90 = 455.
> 2. Divide the sum by the number of values: 455 ÷ 5 = 91. The mean grade is 91.

*Mean, median, and mode are measures of central tendency. The mean is often called the **average**.*

The **median** is the middle value in a set of **ordered data.**

> What is the median of Rowan's grades?
>
> 1. Write the values in order from least to greatest: 89, 89, 90, 92, 95.
> 2. Identify the middle value: 90. The median is 90.

If a data set contains an even number of data values, the median is the average of the middle two values.

The **mode** is the number that appears most often in a data set.

> What is the mode of Rowan's grades?
>
> 1. Write the values in order: 89, 89, 90, 92, 95.
> 2. Identify the number that appears most often: 89. The mode is 89.

A set can have 1 mode, more than 1 mode, or no mode.

The **range** of a data set is the difference between the smallest and largest data values.

> What is the range in Rowan's grades?
>
> 1. Identify the smallest and largest numbers in the data set: 89 is the smallest and 95 is the largest.
> 2. Subtract: 95 − 89 = 6. The range of the grades is 6.

Range is a measure of spread, or distribution.

152 UNIT 7 Statistics

© The Continental Press, Inc. Do not duplicate.

GUIDED PRACTICE

Try this sample multiple-choice problem.

S The ages of the children in the Miller family are listed below.

8, 8, 11, 15, 16, 18

What is the median age of the children?

- A 8
- B 10
- C 13
- D 15

> This problem asks you to find the median of the ages. First make sure the numbers are arranged in order from least to greatest. Since there is an even number of ages, the median is the average of the middle two values, 11 and 15: (11 + 15) ÷ 2 = 26 ÷ 2 = 13. The correct answer is C.

INDEPENDENT PRACTICE

Read each problem. Circle the letter of the best answer.

Use this table to answer questions 1 and 2.

PARKING AMOUNTS HECTOR PAID LAST WEEK

Day	Amount Paid ($)
Monday	2.25
Tuesday	4.75
Wednesday	4.50
Thursday	4.75
Friday	3.50

1 What is the range in dollar amounts Hector paid for parking?

- A $1.25
- B $2.50
- C $4.50
- D $4.75

2 What is the mode amount Hector paid for parking?

- A $3.50
- B $3.95
- C $4.50
- D $4.75

3 The costs of some newspapers are $1.75, $2.50, $3.50, $3.25, and $4.25. What is the median cost of these newspapers?

- A $2.50
- B $3.00
- C $3.25
- D $3.50

4 Margot's homework grades are 78, 85, 99, 70, 88, and 94. If Margot removes her lowest grade, which statement about the range of grades is true?

- A It increases by 5.
- B It decreases by 5.
- C It increases by 8.
- D It decreases by 8.

5 What is the mean of the numbers in the data set below?

6 8 6 11 9 6 5 13

- A 6
- B 7
- C 8
- D 9

UNIT 7 Statistics

GUIDED PRACTICE

Try this sample constructed response problem.

S The numbers of points some players on a basketball team scored during the last game are listed below.

    4   10   13   19   23

Part A: What is the mode number of points scored?

    Answer: __There is no mode.__

Part B: What is the mean number of points scored?

    Answer: __13.8__

> Part A asks you to identify the mode of the data. The mode is the number that appears most often in a data set. Each data value listed appears only once. So this data set has no mode. The correct answer is there is no mode. Part B asks you to find the mean of the data values. To do this, first add all the data values: $4 + 10 + 13 + 19 + 23 = 69$. Then divide the sum by the number of data values: $69 \div 5 = 13.8$. The correct answer is 13.8.

INDEPENDENT PRACTICE

Read the problem. Write your answers.

6 The heights, in feet, of trees Alicia and her friends planted are listed below.

    5.25   4.5   4.75   3   3   3.5

Part A: What is the range, in feet, of tree heights?

    Answer: _____

Part B: What is the mean tree height, in feet?

    Answer: _____

INDEPENDENT PRACTICE

Read the problem. Write your answers.

7 The weights, in pounds, of six pumpkins are listed below.

14   33   24   16   21   24

Part A: What is the median weight, in pounds, of these pumpkins?

Show your work.

Put the data values in order from least to greatest.

Answer: _____

Part B: The weight of a seventh pumpkin is added to this list. The median now becomes the weight of the seventh pumpkin. What could be the weight, in pounds, of the seventh pumpkin?

Answer: _____

Explain how you know.

_____
_____
_____
_____

Bar Graphs

Indicators 6.S.7, 8

graphs   bar graph   predictions   trend

Graphs are used to show data in a visual form. They help make data easier to read and interpret.

A **bar graph** uses bars to show individual pieces of data. Bar graphs are used to compare independent data values.

This bar graph shows the number of minutes Denise used her cell phone the first six months of the year.

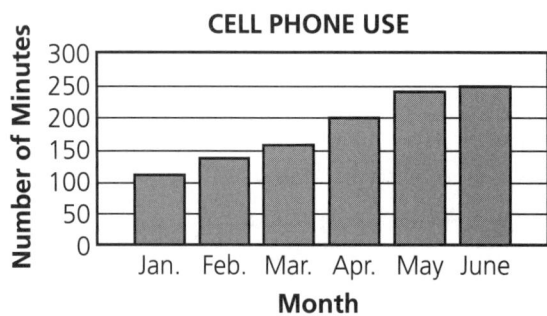

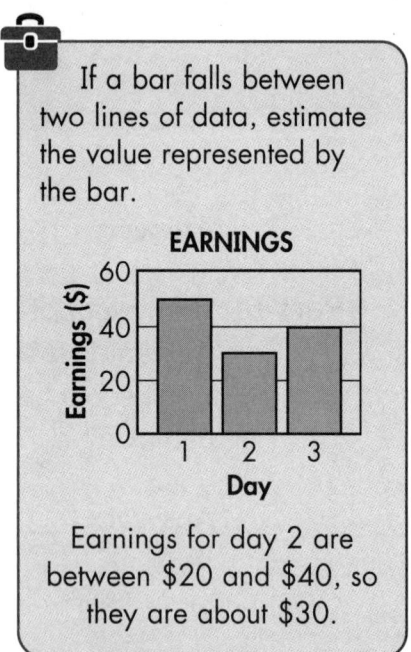

If a bar falls between two lines of data, estimate the value represented by the bar.

Earnings for day 2 are between $20 and $40, so they are about $30.

How many more minutes did Denise use the cell phone in June than April?

1. Find the number of minutes for June and for April: June has 250 minutes and April has 200 minutes.

2. Subtract: 250 − 200 = 50. She used the phone 50 more minutes in June.

You can sometimes use bar graphs to make **predictions** about data.

Based on the data in the bar graph above, what prediction can be made regarding the minutes Denise will use her cell phone in July?

A prediction is a guess about what will happen in the future, based on evidence.

1. Look for a **trend,** or pattern. The number of minutes the cell phone is used increases each month.

2. Make a prediction. Denise will likely use her cell phone more than 250 minutes in July.

UNIT 7 Statistics

© The Continental Press, Inc.   Do not duplicate.

GUIDED PRACTICE

Try this sample multiple-choice problem.

S According to this graph, what fraction of people rented exactly 2 movies this week?

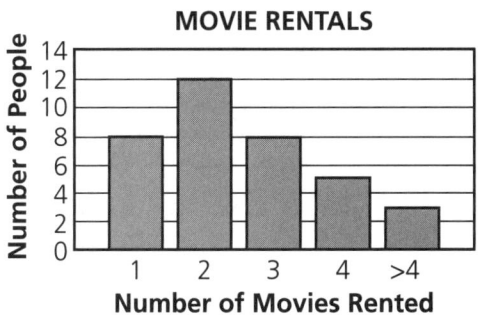

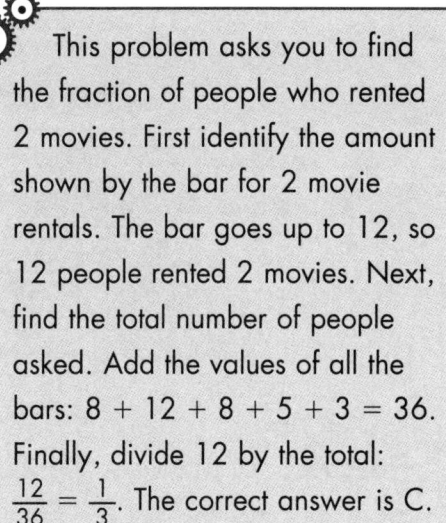

This problem asks you to find the fraction of people who rented 2 movies. First identify the amount shown by the bar for 2 movie rentals. The bar goes up to 12, so 12 people rented 2 movies. Next, find the total number of people asked. Add the values of all the bars: 8 + 12 + 8 + 5 + 3 = 36. Finally, divide 12 by the total: $\frac{12}{36} = \frac{1}{3}$. The correct answer is C.

- **A** $\frac{1}{5}$
- **B** $\frac{1}{4}$
- **C** $\frac{1}{3}$
- **D** $\frac{1}{2}$

INDEPENDENT PRACTICE

Read each problem. Circle the letter of the best answer.

Use this bar graph to answer questions 1 and 2.

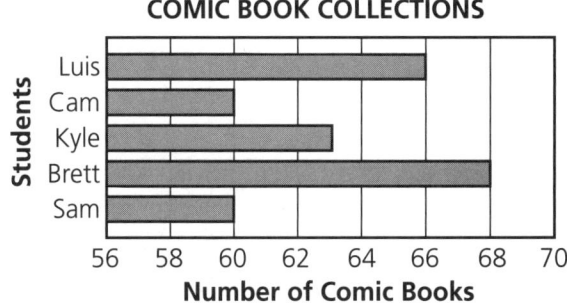

Use this bar graph to answer questions 3 and 4.

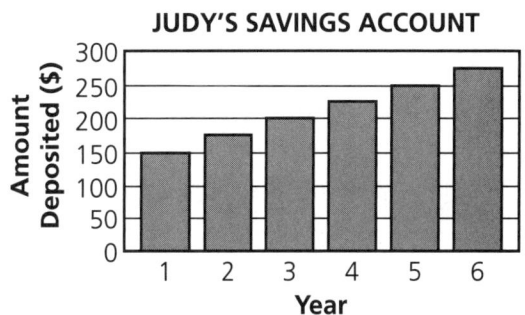

1 Which two students have the same number of comic books?

- **A** Luis and Brett
- **B** Cam and Kyle
- **C** Cam and Sam
- **D** Kyle and Brett

2 How many comic books does Kyle have?

- **A** 62
- **B** 63
- **C** 64
- **D** 65

3 About how much total money did Judy deposit in the first three years?

- **A** $525
- **B** $550
- **C** $575
- **D** $600

4 If this trend continues, in what year will Judy deposit $350 into the account?

- **A** year 6
- **B** year 7
- **C** year 8
- **D** year 9

UNIT 7 Statistics

GUIDED PRACTICE

Try this sample constructed response problem.

S This bar graph shows the prices per night for standard hotel rooms at four different hotels.

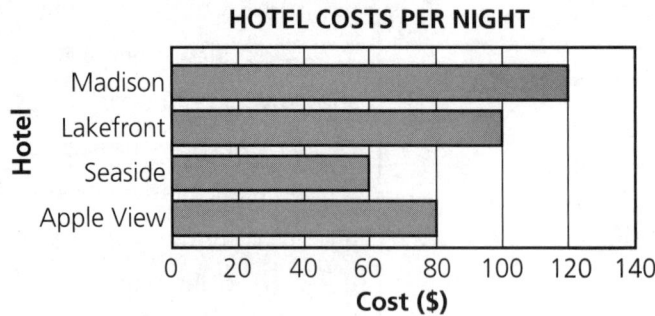

> Part A asks you to identify the hotel that costs $20 more per night than Apple View. The bar for Apple View shows the cost is $80 per night. Find the bar with a cost of $80 + $20, or $100, per night. The correct answer is Lakefront Hotel. Part B asks you to find the cost for 3 nights at Madison Hotel. The bar for Madison shows each night is $120, so 3 × $120 = $360. The correct answer is $360.

Part A: Which hotel costs $20 more per night than Apple View Hotel?

Answer: ___Lakefront Hotel___

Part B: Cristina plans to stay in a standard room at Madison Hotel for 3 nights. What will be her total cost?

Answer: ___$360___

INDEPENDENT PRACTICE

Read the problem. Write your answers.

5 The number of hours Lindsey worked at the mall each day last week is shown in this bar graph.

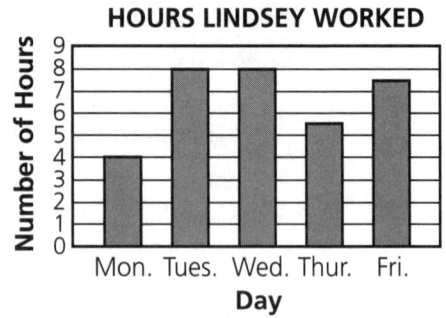

Part A: Which day did Lindsey work exactly $5\frac{1}{2}$ hours?

Answer: _____

Part B: How many total hours did Lindsey work last week?

Answer: _____

UNIT 7 Statistics

INDEPENDENT PRACTICE

Read the problem. Write your answers.

6 The number of magazine subscriptions Hiro sold during four weeks is shown in this bar graph.

Part A: How many more subscriptions did Hiro sell in week 1 than in week 2?

Answer: _____

> A bar that falls between two lines stands for a number between the two numbers shown.

Part B: Hiro has a goal to sell 50 magazine subscriptions. Based on the trend in this graph, do you think he will reach his goal in week 5?

Answer: _____

Explain how you know.

_____
_____
_____
_____

Histograms

Indicators 6.S.7, 8

histogram   interval

A **histogram** uses bars to show frequency data. Each bar stands for one **interval.** Unlike a bar graph, the bars in a histogram are side-by-side since the intervals are continuous.

Bars in a histogram touch each other. Bars in a bar graph do not.

About how many customers are served between 11 and 1?

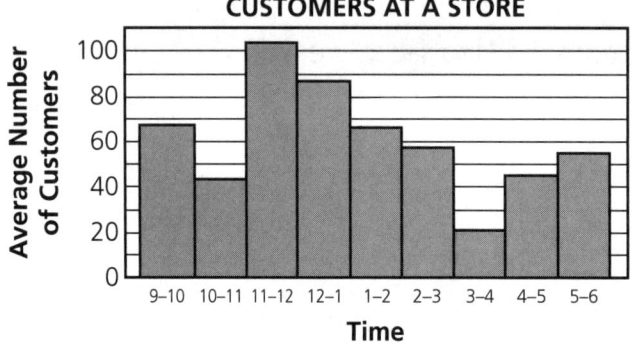

The interval is the difference between the numbers on the scale.

1. Read the value of the bars: there are about 105 customers between 11 and 12 and 88 customers between 12 and 1.

2. Add the values: 105 + 85 = 180. There are about 180 customers between 11 and 1.

Histograms can also be used to make conclusions about data.

Kasey wants to go shopping when the store is not too busy. What time would be best for her to choose?

There should be no gaps in the intervals shown on a histogram.

1. Decide what information to look for. A store is not busy when there are few customers. This would be shown by a short bar.

2. Look for the time interval with the shortest bar. The shortest bar is for the time from 3 and 4.

GUIDED PRACTICE

Try this sample multiple-choice problem.

S About how much was the total water bill for the year?

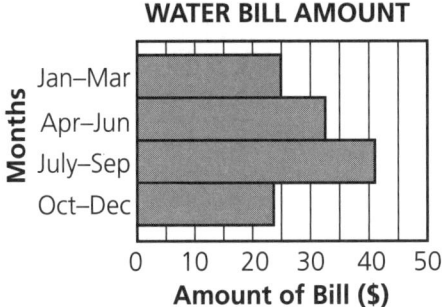

> This problem asks you to find the total water bill amount for the year. First find the approximate values represented by each bar. Then add them together. The total is about $25 + $33 + $41 + $23 = $122. The correct answer is B.

A $116 C $128
B $122 D $134

INDEPENDENT PRACTICE

Read each problem. Circle the letter of the best answer.

Use this histogram to answer questions 1 and 2.

Use this histogram to answer questions 3 and 4.

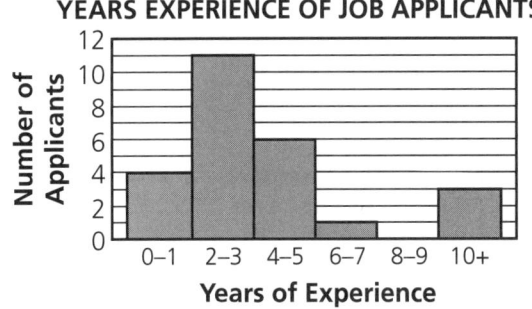
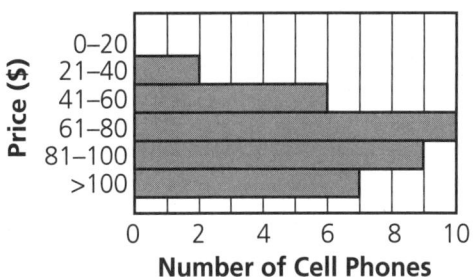

1 How many applicants have between 4 and 5 years experience?

A 4   C 6
B 5   D 7

2 Another job applicant is included in the data. How many years experience is this person **most likely** to have?

A 1   C 5
B 3   D 10

3 How many cell phones sold that day were priced more than $80?

A 7   C 16
B 10   D 19

4 Which statement can be concluded from this graph?

A No cell phones sold for $20.
B Two cell phones sold for $30.
C Five cell phones sold for $70.
D No cell phones sold for $200.

UNIT 7 Statistics

GUIDED PRACTICE

Try this sample constructed response problem.

S This histogram shows the results of a survey on the year of people's car models.

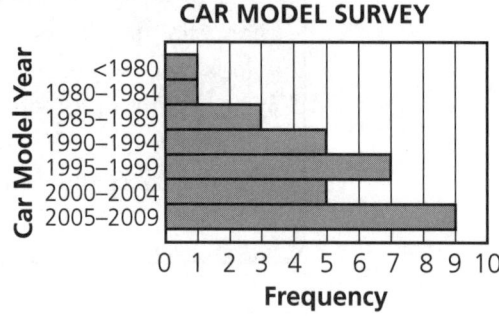

> Part A asks you to find the number of responders with car models from the years 1995 to 1999. The height for this bar reaches to 7, so the correct answer is 7. Part B asks you to find the number of survey responders with car models newer than 1999. To do this, add the bar heights for the intervals 2000–2004 and 2005–2009: 5 + 9 = 14. The correct answer is 14.

Part A: How many survey responders have a car model from the years 1995 to 1999?

Answer: _____7_____

Part B: How many survey responders have a car model newer than 1999?

Answer: _____14_____

INDEPENDENT PRACTICE

Read the problem. Write your answers.

5 The number of Internet orders a business received during June is shown in this histogram.

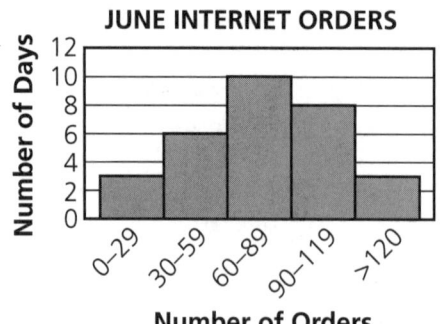

Part A: On how many days in June did the business receive fewer than 60 Internet orders?

Answer: _____

Part B: What fraction of the month had between 60 and 89 Internet orders?

Answer: _____

INDEPENDENT PRACTICE

Read the problem. Write your answers.

6 This histogram shows the number of points students in a class earned in a review game.

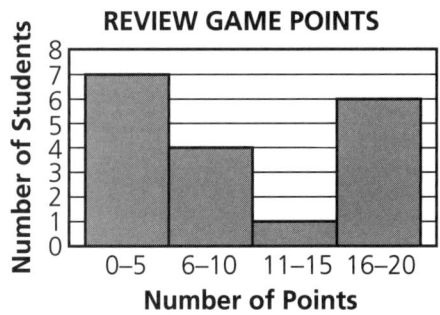

Part A: How many students participated in the game?

Answer: _____

Part B: One-third of the students got bonus points in the review game. What conclusion can be made about these students based on the data in the histogram?

Find the interval that is made up of one-third of the students.

_____
_____
_____
_____

UNIT 7 Statistics

Lesson 4: Line Graphs

Indicators 6.S.7, 8

line graph   scale   interval   trend

A **line graph** shows data over time. The points on the graph show the data value for one moment in time, such as an hour, a month, or a year. A broken line is drawn to connect each point.

> The **scale** is the numbers on the side or the bottom of a graph.

This line graph shows the number of total miles on a car at the beginning of each year.

About how many total miles were on the car at the beginning of year 3?

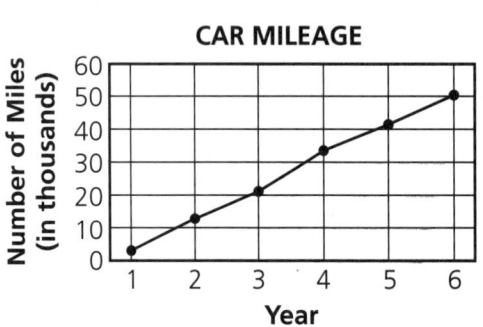

1. Find year 3 on the horizontal scale and move up until you reach the point on the graph.

> The **interval** is the difference between the numbers on the scale.

2. Look to the left to see what number of miles this point corresponds to. The point at year 3 is close to 20,000 miles.

You can use line graphs to interpret data. The direction of a line on a graph shows the **trend** of the data. A line that moves upward shows an increasing trend. A line that moves downward shows a decreasing trend.

> When a point falls between two numbers on the scale, it stands for a number in-between. You can estimate the value of the point based on where it is between the numbers on the scale.

How does the mileage on the car change each year?

1. Look for a trend in the data. The line moves upward about the same amount from year to year. So there is an increasing trend.

2. Find the difference between points that are next to each other. Between years 5 and 6, the mileage increases about 50,000 − 40,000 = 10,000 miles. So the car is driven about 10,000 miles each year.

UNIT 7 Statistics

GUIDED PRACTICE

Try this sample multiple-choice problem.

S About how much more money was in the checking account on April 1st than on April 15th?

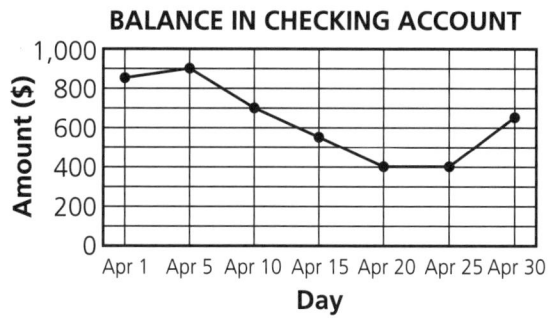

- **A** $300
- **B** $350
- **C** $400
- **D** $450

> This problem asks you to find the difference between amounts on April 1 and April 15. The dollar amount shown on April 1 is between $800 and $900, so it is around $850. The dollar amount shown on April 15 is between $500 and $600, so it is around $550. The difference between these amounts is $850 − $550 = $300. The correct answer is A.

INDEPENDENT PRACTICE

Read each problem. Circle the letter of the best answer.

Use this line graph to answer questions 1 and 2.

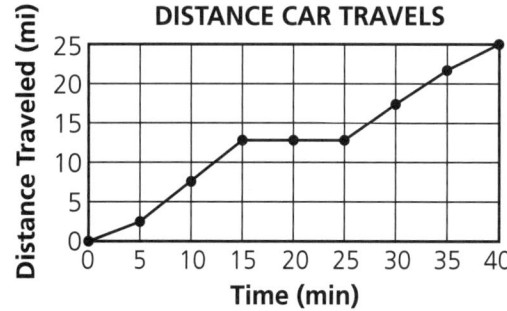

1 About how far did the car travel in the first 10 minutes?

- **A** 7 mi
- **B** 10 mi
- **C** 12 mi
- **D** 15 mi

2 The car was stopped in traffic during its trip. How many minutes was it stopped?

- **A** 5
- **B** 10
- **C** 15
- **D** 25

Use this line graph to answer questions 3 and 4.

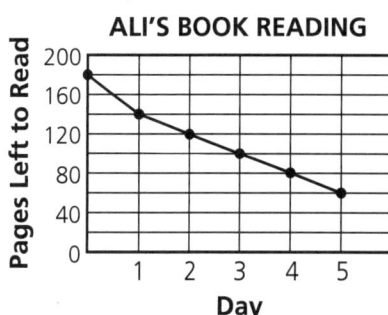

3 How many pages did Ali read the first day?

- **A** 20
- **B** 40
- **C** 140
- **D** 180

4 If Ali continues reading at the same rate, on which day would she expect to finish the book?

- **A** day 6
- **B** day 7
- **C** day 8
- **D** day 9

UNIT 7 Statistics

GUIDED PRACTICE

Try this sample constructed response problem.

S This line graph shows how the price per game of bowling changed each year.

> Part A asks you to find the difference in price between years 2 and 3. The price in year 2 was $2.50. The price in year 3 was $3.00. The difference in prices is $3.00 − $2.50 = $0.50. Part B asks you to find the total cost for bowling 4 games at the year 5 price. The price in year 5 is $3.50 per game. So the cost for 4 games is 4 × $3.50 = $14. The correct answer is $14.00.

Part A: How much did the price per game increase from year 2 to year 3?

Answer: ____$0.50____

Part B: Willie bowled 4 games at the price shown in year 5. What was his total cost for these games?

Answer: ____$14.00____

INDEPENDENT PRACTICE

Read the problem. Write your answers.

5 This line graph shows the salary of a clerk at a music store.

Part A: How much more was the salary of this clerk in year 6 than in year 1?

Answer: _____

Part B: Between which two years did the amount of the clerk's salary increase the most?

Answer: _____

UNIT 7 Statistics

Read the problem. Write your answers.

6 This line graph shows the amount of money a company budgeted for expenses each quarter this year.

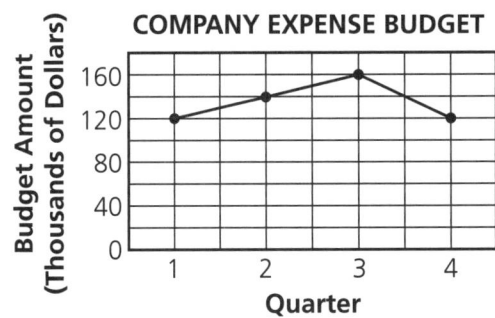

Part A: How much money, in thousands of dollars, was budgeted for all four quarters?

Answer: _____

Part B: Every year, the company budgets money for expenses following the trend shown in this line plot. Next year, the amount budgeted for each quarter will be $\frac{4}{5}$ the amount budgeted this year. How much money will be budgeted for expenses in quarter 3 next year?

Show your work.

Find the amount budgeted in quarter 3 for this year. Decide how it will change next year.

Answer: _____

UNIT 7 Statistics

Circle Graphs

Indicators 6.S.7, 8

circle graph   pie chart

A **circle graph** is a data display that shows the parts of a whole, or 100%. Each section represents a part of the entire amount.

> Another name for a circle graph is a **pie chart**.

This circle graph shows the first letter of students' last names in Derek's class.

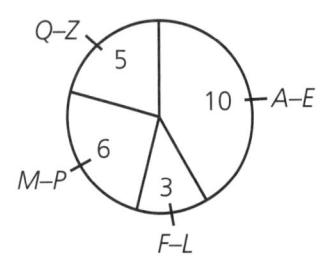

LAST NAME INITIALS

> The labels on a circle graph may be whole numbers, decimals, fractions, or percents.

What fraction of students have a last name beginning with the letters A–E?

1. Find the number of students with a last name beginning with A–E: 10.

2. Find the total number of students in the class by adding each amount: 10 + 3 + 6 + 5 = 24 total students.

3. Divide the number of A–E students by the total students and simplify: $\frac{10}{24} = \frac{10 \div 2}{24 \div 2} = \frac{5}{12}$. So $\frac{5}{12}$ of the students have a last name beginning with A–E.

> To change a percent to a decimal, drop the percent sign and move the decimal point two places to the left.

UNIT 7 Statistics

GUIDED PRACTICE

Try this sample multiple-choice problem.

S A total of 160 dogs were at a dog show.

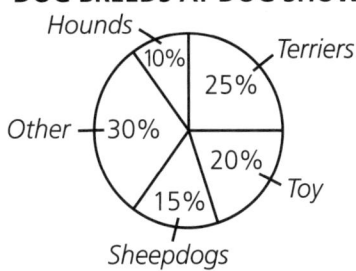

How many of these dogs were sheepdogs?

- **A** 15
- **B** 18
- **C** 24
- **D** 30

> This problem asks you to find the number of sheepdogs at a dog show. There are 160 dogs total at the dog show and 15% of them are sheepdogs. To find the number of sheepdogs, multiply: 160 × 15% = 160 × 0.15 = 24. The correct answer is C.

INDEPENDENT PRACTICE

Read each problem. Circle the letter of the best answer.

Use this circle graph to answer questions 1 and 2.

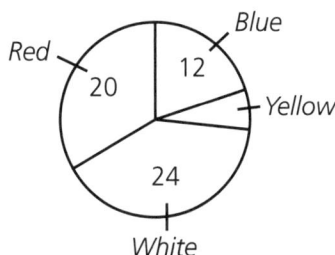

1 A total of 60 T-shirts were sold. How many were yellow?

- **A** 4
- **B** 6
- **C** 8
- **D** 10

2 Which color made up 20% of the 60 sold T-shirts?

- **A** red
- **B** blue
- **C** white
- **D** yellow

Use this circle graph to answer questions 3 and 4.

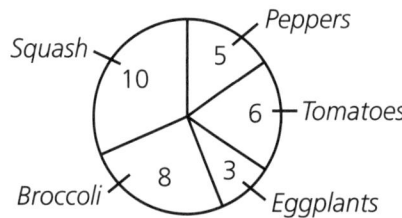

3 What percent of the vegetable plants bought are broccoli plants?

- **A** 8%
- **B** 20%
- **C** 25%
- **D** 40%

4 If each tomato plant produces an average of 50 to 60 pounds of tomatoes, about how many pounds are all the tomato plants expected to produce?

- **A** 300
- **B** 400
- **C** 500
- **D** 600

UNIT 7 Statistics

GUIDED PRACTICE

Try this sample constructed response problem.

S The fraction of children enrolled in each weekly session of a summer camp is shown in the circle graph.

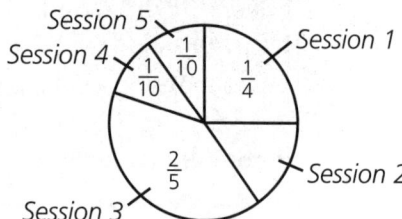

STUDENTS ENROLLED IN SUMMER CAMP

> Part A asks you to find the fraction of children enrolled in session 2. Add the given fractions and subtract the sum from 1 whole: $\frac{1}{4} + \frac{2}{5} + \frac{1}{10} + \frac{1}{10} = \frac{5}{20} + \frac{8}{20} + \frac{2}{20} + \frac{2}{20} = \frac{17}{20}$ and $1 - \frac{17}{20} = \frac{20}{20} - \frac{17}{20} = \frac{3}{20}$. The correct answer is $\frac{3}{20}$. Part B asks you to find how many more children are in session 1 than session 4. There are $300 \times \frac{1}{4} = 75$ children in session 1 and $300 \times \frac{1}{10} = 30$ children in session 4. The difference is $75 - 30 = 45$. The correct answer is 45.

Part A: What fraction of children are enrolled in session 2 of the summer camp?

Answer: $\frac{3}{20}$

Part B: A total of 300 children are enrolled in the five weekly sessions. How many more children are enrolled in session 1 than in session 4?

Answer: 45

INDEPENDENT PRACTICE

Read the problem. Write your answers.

5 This circle graph shows the types of beverages 80 students had with lunch today.

What percent of students had milk with lunch?

Show your work.

LUNCH BEVERAGES

Answer: _____

170 UNIT 7 Statistics

INDEPENDENT PRACTICE

Read the problem. Write your answers.

6 Regan surveyed customers at a florist shop about their favorite rose color. The results of her survey are shown in the circle graph below.

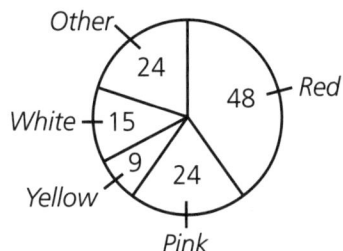

FAVORITE ROSE COLOR

Part A: What fraction of people surveyed said red was their favorite color rose?

Show your work.

Find the total number of people surveyed.

Answer: _____

Part B: A florist plans to order 600 dozen roses. Based on the results of this survey, how many dozen yellow roses should Regan recommend the florist order?

Answer: _____

Explain your reason.

_____
_____
_____
_____

UNIT 7 Statistics

Statistics Review

Read each problem. Circle the letter of the best answer.

Use this bar graph to answer questions 1–4.

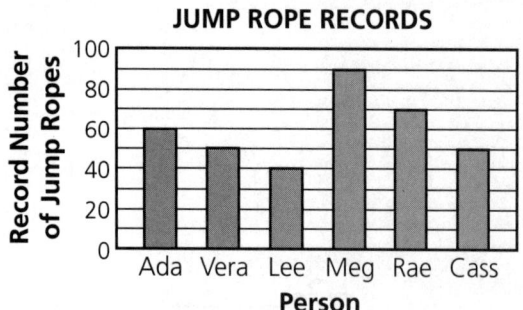

1 What is the mode of this data?

   A 50   **C** 60

   B 55   **D** 80

2 How many more times did Rae jump rope than Vera?

   A 20   **C** 40

   B 30   **D** 50

3 What is the range in record number of jumps?

   A 40   **C** 60

   B 50   **D** 100

4 What is the median number of record jumps?

   A 50

   B 55

   C 60

   D 65

Use this line graph to answer questions 5 and 6.

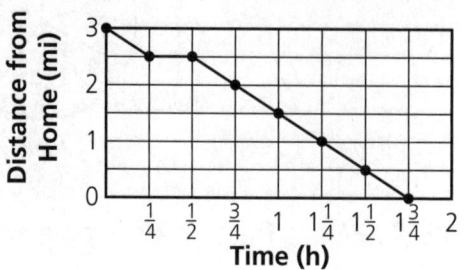

5 How far is Jared from home after 1 hour?

   A 1 mi   **C** $1\frac{1}{2}$ mi

   B $1\frac{1}{4}$ mi   **D** 2 mi

6 When Jared is between 1 and 2 miles from home, what is his average walking speed, in miles per hour?

   A $\frac{1}{4}$ mph

   B $\frac{1}{2}$ mph

   C 1 mph

   D 2 mph

7 The student populations of five schools in one district are as follows: 219, 343, 278, 409, and 306. What is the mean school population?

   A 263   **C** 306

   B 278   **D** 311

Statistics Review

Read each problem. Write your answers.

8 This histogram shows the finish times of runners in a race.

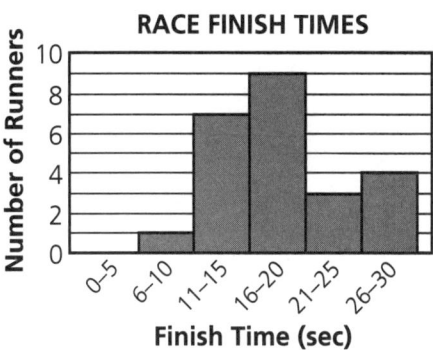

   Part A: How many runners finished the race in more than 10 seconds but less than 21 seconds?

   Answer: _____

   Part B: What fraction of runners finished the race in more than 25 seconds?

   Answer: _____

9 This circle graph shows the favorite subjects of the 25 students in homeroom A.

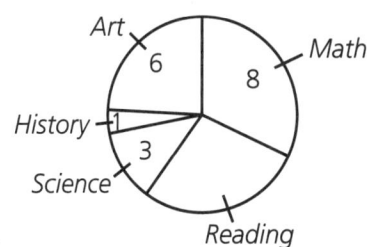

   Part A: How many of these students said reading was their favorite subject?

   Answer: _____

   Part B: Twice as many of the 25 students in homeroom B said math was their favorite subject. What percent of these students said math was their favorite subject?

   Answer: _____

Read the problem. Write your answers.

10 The scores each gymnast on one team received at a sports meet are listed below.

      9.7   8.7   7.9   9.5   9.0   9.2

Part A: What is the range of these scores?

Answer: _____

Part B: Which is greater, the mean or the median of these scores? Show your work.

Answer: _____

UNIT 8

Probability

Probability is the chance that something will happen. You use it when you decide what your chances are of pulling the card you need to win a game. Part of understanding probability is finding the possible outcomes for different events. For example, when you choose a book at random from a bookshelf, each book is a possible outcome. You can use probability to make predictions about what might happen. When you are practicing shooting a basketball in your driveway, you might make 15 baskets out of 30 shots. So the chance of you making a basket is $\frac{1}{2}$. You can predict that if you shoot 60 more times, you will make 30 more baskets.

This unit will help you answer test questions about probability. There are three lessons in this unit:

1. Probability   In this lesson, you will review how to find probability for different events. You will write probability as a ratio, a fraction, or with words.

2. Sample Spaces   This lesson reviews how to find the sample space for an event. You will show a sample space as an organized list and as a tree diagram.

3. Predictions from Data   This lesson reviews how to make predictions using a sample of data and probability.

Probability

Indicators 5.S.5, 6

probability   event   favorable outcomes   possible outcomes

Probability is the chance that a certain **event,** or outcome, will happen. Probabilities can be written as ratios that compare the number of **favorable outcomes** to the number of **possible outcomes.**

Justina has these shapes.

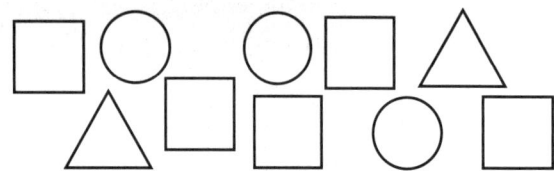

She wants to know what her chances are of picking a circle from these shapes without looking. What is the ratio of favorable outcomes to possible outcomes?

1. Identify the number of favorable outcomes. Picking a circle is the favorable outcome. There are 3 circles. So the number of favorable outcomes is 3.

2. Identify the number of possible outcomes, or total shapes. There are 5 squares, 2 triangles, and 3 circles. In all, there are 5 + 2 + 3 = 10 shapes. So the number of possible outcomes is 10.

3. Write the ratio of favorable outcomes to possible outcomes: 3 out of 10 or 3:10.

A favorable outcome is what you are looking for. Possible outcomes are the total number of outcomes.

Probability =
$P(\text{event}) = \dfrac{\text{Favorable outcomes}}{\text{Possible outcomes}}$

Probabilities can also be written as fractions.

What is the probability Justina will pick a square?

1. Identify the number of favorable outcomes: 5 squares.

2. Identify the number of possible outcomes: 10 total shapes.

3. Write the probability as a fraction: $\dfrac{5}{10}$ or $\dfrac{1}{2}$.

When a probability is written as a fraction, simplify it to lowest terms.

$\dfrac{6}{8}$ simplifies to $\dfrac{3}{4}$ or 3:4 or 3 out of 4.

UNIT 8 Probability

GUIDED PRACTICE

Try this sample multiple-choice problem.

S What is the probability of randomly picking a vowel from this set of balls?

- **A** 1 out of 2
- **B** 1 out of 5
- **C** 2 out of 3
- **D** 2 out of 5

> This problem asks you to find the probability of picking a ball with a vowel from a set. There are 2 favorable outcomes, or vowels: A and E. There are 5 possible outcomes, or total letters: A, B, C, D, and E. So, the probability is 2 out of 5. The correct answer is D.

INDEPENDENT PRACTICE

Read each problem. Circle the letter of the best answer.

1 The seasons of the year are written on cards and placed faced down. A card is picked at random. What are the possible outcomes?

- **A** winter and summer
- **B** spring and summer
- **C** winter, spring, and summer
- **D** winter, spring, summer, and fall

2 Enrico rolls a number cube with the numbers 1 through 6 on it. What is the probability that he rolls the number 3?

- **A** 1:2
- **B** 1:3
- **C** 1:6
- **D** 3:6

3 In a class of 21 students, 9 are girls and the rest are boys. What is the probability that a boy will be chosen at random from these students?

- **A** 1 out of 9
- **B** 3 out of 7
- **C** 4 out of 7
- **D** 4 out of 9

4 Tori wants to know the probability of picking an even number from this set.

$$S = \{2, 4, 5, 7, 8, 9\}$$

What are the favorable outcomes?

- **A** 2 and 4 only
- **B** 2, 4, and 8 only
- **C** 5, 7, and 9 only
- **D** 2, 4, 5, 7, 8, 9

5 These numbers are on slips of paper.

| 2,012 | 2,210 | 1,021 |
| 1,022 | 2,010 | 1,200 |

Morgan will pick one at random. What is the probability that she picks a number with a 2 in the hundreds place?

- **A** $\frac{1}{6}$
- **B** $\frac{1}{3}$
- **C** $\frac{1}{2}$
- **D** $\frac{2}{3}$

UNIT 8 Probability

GUIDED PRACTICE

Try this sample constructed response problem.

S Brie's cabinet has 6 cans of vegetable soup, 4 cans of chicken soup, and 8 cans of beef barley soup. She will pick a can of soup from the cabinet at random.

Part A: What is the probability Brie will pick a can of vegetable soup? Write your answer in ratio form.

Answer: ____6:18____

Part B: What is the probability of picking a can that is **not** chicken soup? Write your answer as a fraction in lowest terms.

Answer: ____$\frac{7}{9}$____

> Part A asks you to find the probability of picking vegetable soup. The number of favorable outcomes, or cans of vegetable soup, is 6. The number of possible outcomes, or total cans of soup, is $6 + 4 + 8 = 18$. As a ratio, the correct answer is 6:18 or 1:3. Part B asks you to find the probability of picking a can that is not chicken soup. The favorable outcomes are vegetable soup and beef barley soup. There are $6 + 8 = 14$ favorable outcomes. So, the probability is $\frac{14}{18} = \frac{7}{9}$. The correct answer is $\frac{7}{9}$.

INDEPENDENT PRACTICE

Read the problem. Write your answers.

6 Ted has 5 white mugs, 4 blue mugs, 1 yellow mug, and 6 brown mugs. He picks a mug without looking.

Part A: How many possible outcomes are there for him to pick a mug?

Answer: _____

Part B: What is the probability he picks a blue or a brown mug? Write your answer as a fraction in lowest terms.

Answer: _____

INDEPENDENT PRACTICE

Read the problem. Write your answers.

7 Danuta is trying to guess what month Angela has her birthday. Angela told Danuta her birthday is in a month that has the letter R in it.

Part A: List all possible outcomes for the birthday month.

Answer: _____

_____

Part B: What is the probability that the month ends in the letter R? Write your answer as a fraction in lowest terms.

Show your work.

> Find the favorable outcomes of the months with the letter R.

Answer: _____

Sample Spaces

Indicator 5.S.7

sample space   organized list   fair coin

A **sample space** shows all the possible outcomes for an event. An **organized list** can be used to show the sample space.

What is the sample space for rolling a 1–6 number cube?

1. Find the possible outcomes for the event. A 1–6 number cube has six sides marked with the numbers from 1 through 6.

2. List the possible outcomes in order: 1, 2, 3, 4, 5, 6. The sample space is the numbers 1, 2, 3, 4, 5, 6.

> The possible outcomes of an event are all the outcomes. The favorable outcomes are the ones you are looking for from the possible outcomes.

You can use sample spaces to determine the probability of an event.

Tatiana rolls a 1–6 number cube. What is the probability the number cube lands on an odd number?

1. Find the number of possible outcomes. The sample space above shows there are 6 possible outcomes.

2. Find the number of favorable outcomes. There are 3 favorable outcomes, or odd numbers: 1, 3, and 5.

3. Find the probability.

$$\text{Probability} = \frac{\text{Favorable outcomes}}{\text{Possible outcomes}} = \frac{3}{6} = \frac{1}{2}$$

The probability of rolling an odd number is $\frac{1}{2}$.

> Probability =
> $P(\text{event}) = \frac{\text{Favorable outcomes}}{\text{Possible outcomes}}$

> A **fair coin** is one that has an equally likely chance of landing on heads or tails.

UNIT 8 Probability

GUIDED PRACTICE

Try this sample multiple-choice problem.

S A cooler has 2 cans of cola (C), 3 cans of root beer (R), and 1 can of ginger ale (G). Which list shows all the possible outcomes of cans in the cooler?

A C, R, G

B C, C, R, R, G, G

C C, C, R, R, R, G

D C, C, C, R, R, R, G, G, G

> This problem asks you to find the total possible outcomes of cans in a cooler. Make an organized list. Two cans of cola are represented as C, C. Three cans of root beer are represented as R, R, R. One can of ginger ale is represented as G. Combine these to get C, C, R, R, R, G. The correct answer is C.

INDEPENDENT PRACTICE

Read each problem. Circle the letter of the best answer.

1 What are the possible outcomes to flip a fair coin with heads (H) and tails (T)?

A T
B H
C H, T
D H, T, H, T

Use these coins to answer questions 2 and 3.

2 How many items are in the sample space for these coins?

A 1
B 2
C 4
D 8

3 What is the probability that a coin picked at random will be a nickel?

A $\frac{1}{8}$
B $\frac{1}{4}$
C $\frac{1}{3}$
D $\frac{1}{2}$

4 Anita listed the house numbers on her street.

16 25 49 51 63 74 81 82

She put the tens digits from each number in a sample space. Which list shows Anita's sample space?

A 1, 2, 4, 5, 6, 7, 8, 8
B 1, 1, 2, 3, 4, 5, 6, 9
C 1, 2, 3, 4, 5, 6, 7, 8
D 1, 2, 3, 4, 5, 6, 7, 8, 9

5 Carl has 6 pairs of white socks, 2 pairs of blue socks, and 4 pairs of black socks. He picks a pair without looking. What is the probability that the socks are white?

A $\frac{1}{2}$
B $\frac{1}{3}$
C $\frac{1}{4}$
D $\frac{1}{6}$

UNIT 8 Probability

GUIDED PRACTICE

Try this sample constructed response problem.

S   Franklyn spins the arrow on this spinner.

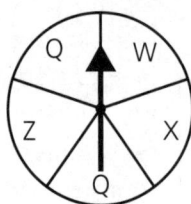

> Part A asks you to write all possible outcomes for the spinner. The possible outcomes are the letters in each section of the spinner: Q, Q, W, X, and Z. Since the letter Q appears twice on the spinner, it is listed twice as a possible outcome. Part B asks you to find the probability of the arrow landing on either Q or W. From the 5 possible outcomes listed in part A, 2 outcomes are Q and 1 outcome is W. So, there are 3 favorable outcomes altogether. The probability is $\frac{3}{5}$.

Part A: List all possible outcomes for this spinner.

Answer: ___Q, Q, W, X, Z___

Part B: What is the probability that the arrow will land on Q or W?

Answer: ___$\frac{3}{5}$___

INDEPENDENT PRACTICE

Read the problem. Write your answers.

6   A cooler holds 6 tuna (T) sandwiches, 4 ham (H) sandwiches, and 2 egg salad (E) sandwiches. Bindi will pick a sandwich at random.

Part A: Make a list to show all the possible outcomes for Bindi's selection.

Answer: _____

Part B: What is the probability that Bindi will pick a tuna or egg salad sandwich?

Answer: _____

UNIT 8 Probability

INDEPENDENT PRACTICE

Read the problem. Write your answers.

7 A cell phone case is given away as a prize at a carnival game. There are 4 white (W) cases, 9 black (B) cases, 3 red (R) cases, 2 yellow (Y) cases, 2 orange (O) cases, and 4 green (G) cases. Game winners choose a cell phone case at random.

Part A: Make an organized list to show all the possible outcomes for picking a cell phone case.

Answer: _____

How many possible options are there?

Answer: _____

Part B: What is the probability of selecting a cell phone case at random that is **not** red or white?

Answer: _____

> Count the number favorable outcomes. Count the number of total outcomes.

Lesson 3: Predictions from Data

Indicator 6.S.8

prediction   sample   population   sample probability

You can collect data and make a **prediction** from it when you do not know all the possible outcomes of an event, or how likely each outcome is. A **sample** of data can be used to determine probability.

> A sample is a representative group from an entire **population**.

Jodie asked a group of 50 students from her school if they buy hot lunch every day. Of them, 35 said they did. Based on the survey results, how many of the 600 students in the school would be expected to buy hot lunch every day?

1. Find the probability from the sample of 50.

$$P(\text{hot lunch}) = \frac{\text{Favorable outcomes}}{\text{Possible outcomes}} = \frac{35}{50} = \frac{7}{10}$$

> Probability =
> $P(\text{event}) = \frac{\text{Favorable outcomes}}{\text{Possible outcomes}}$

2. Multiply the **sample probability** by the population of 600: $\frac{7}{10} \times 600 = 420$. So 420 of the 600 students would be expected to buy hot lunch every day.

You can also set up a proportion to find the predicted probability of an event.

Malachi inspected 20 lampshades from a shipment of 200 and found 3 were defective. Predict the number of lampshades from the entire shipment that can be expected to be defective.

> The larger the sample, the more accurate the predicted probability will be.

1. Set up a proportion.

$$\frac{\text{Defective in sample}}{\text{Sample size}} = \frac{\text{Defective in population}}{\text{Population size}}$$

$$\frac{3}{20} = \frac{n}{100}$$

2. Solve the proportion.

$$20 \times n = 3 \times 200$$
$$20 \times n = 600$$
$$n = 600 \div 20 = 30$$

So 30 lampshades from the entire shipment can be expected to be defective.

184   UNIT 8 Probability

GUIDED PRACTICE

Try this sample multiple-choice problem.

S A box contains 300 colored paper clips. Tony pulled 25 paper clips out at random and found 6 pink, 4 white, 10 yellow, and 5 blue clips in the sample. From the data, Tony predicts there are about 150 yellow paper clips in the entire box. Is Tony's prediction reasonable?

- **A** Yes, it should be **about** 150.
- **B** No, it should be **about** 75.
- **C** No, it should be **about** 120.
- **D** No, it should be **about** 200.

> This problem asks you to decide if a predicted amount is reasonable. First find the sample probability: $\frac{10 \text{ yellow}}{25 \text{ in sample}} = \frac{2}{5}$. Next, multiply the sample probability by the total amount: $\frac{2}{5} \times 300 = 120$. The predicted amount should be close to 120. The correct answer is C.

INDEPENDENT PRACTICE

Read each problem. Circle the letter of the best answer.

1 Ella rolled a 1–6 number cube 15 times. The results of each roll are shown below.

```
1   3   2   1   6
2   5   1   3   5
4   5   3   1   2
```

Based on these results, what is the probability of rolling a 2 the next time?

- **A** $\frac{1}{6}$
- **B** $\frac{1}{5}$
- **C** $\frac{1}{3}$
- **D** $\frac{1}{2}$

2 A store sold 64 sweatshirts one day. Of these, 12 were gray. The store manager plans to order 400 more sweatshirts. Based on the data, how many gray sweatshirts should the manager order?

- **A** 50
- **B** 75
- **C** 100
- **D** 125

3 The arrow is spun on a spinner 100 times. It lands on X 40 times, on Y 20 times, and on Z 40 times. Which spinner is **most likely** to have been spun?

4 In a sample of 24 people at a concert, 15 were under age 20. If 5,593 people attended the concert, is 3,500 a good prediction of the number under 20?

- **A** Yes, it should be **about** 3,500.
- **B** No, it should be **about** 2,500.
- **C** No, it should be **about** 3,000.
- **D** No, it should be **about** 4,000.

UNIT 8 Probability

GUIDED PRACTICE

Try this sample constructed response problem.

S Cole spun the arrow on a spinner 75 times. Red resulted 30 times, blue resulted 10 times, green resulted 20 times, and purple resulted the rest of the times.

Part A: What is the predicted probability that purple will be the result of the next spin of the arrow?

Answer: $\frac{1}{5}$

Part B: Cole will spin the arrow on the spinner a total of 250 times. Predict the number of times the arrow will land on purple.

Answer: 50

> Part A asks you to find the probability that purple will result on the next spin. Find the sample probability, $\frac{\text{favorable outcomes}}{\text{possible outcomes}}$. The possible outcomes are given as 75. The favorable outcomes are 75 − (30 + 10 + 20) = 75 − 60 = 15. The sample probability is $\frac{15}{75} = \frac{1}{5}$. The correct answer is $\frac{1}{5}$. Part B asks you to find the number of times the arrow will land on purple when spun 250 times. Multiply the sample probability by the total number of spins: $\frac{1}{5} \times 250 = 50$. The correct answer is 50 times.

INDEPENDENT PRACTICE

Read the problem. Write your answers.

5 A book club offer was mailed to a sample of 400 families. Of these, 80 families ordered books. Next month, the book club offer will be mailed to 20,000 families. What is the predicted number of orders expected from the mailing next month?

Show your work.

Answer: _____

UNIT 8 Probability

INDEPENDENT PRACTICE

Read the problem. Write your answers.

6 Ava has a bag with 500 coins. It includes only pennies, nickels, dimes, and quarters. She takes 30 coins from the bag and counts them. The results are recorded in the table below.

AVA'S COINS

Pennies	12
Nickels	6
Dimes	2
Quarters	10

Part A: What is the predicted probability that a coin taken from the bag will be a penny?

Answer: _____

Part B: Based on the results in the table, Ava predicts there are 165 quarters in the bag. Is Ava's prediction reasonable?

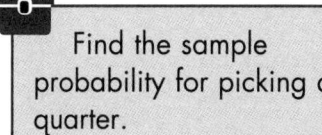

Find the sample probability for picking a quarter.

Answer: _____

Explain how you know.

_____
_____
_____
_____

Probability Review

Read each problem. Circle the letter of the best answer.

1 A hair salon is not open on weekends. What are the possible outcomes for the day a customer can get a haircut?

A Saturday and Sunday

B Monday, Wednesday, Friday

C Monday, Tuesday, Wednesday, Thursday, Friday

D Sunday, Monday, Tuesday, Wednesday, Thursday, Friday, Saturday

2 In a parking lot, there are 5 vans, 8 SUVs, 6 trucks, and 11 cars. What is the probability that the next vehicle to leave the parking lot is a truck?

A $\frac{1}{6}$ C $\frac{1}{4}$

B $\frac{1}{5}$ D $\frac{1}{3}$

3 A spinner contains these numbers.

15	18	21	24	27
30	33	36	39	42

Patty wants to know the probability of getting a number that has an odd digit in the tens place. What are the favorable outcomes?

A 15, 18

B 21, 24, 27, 42

C 15, 21, 27, 33, 39

D 15, 18, 30, 33, 36, 39

4 In a survey of 200 residents, 75 said they support the proposed town budget. If there are 10,000 total residents, about how many would be expected to support the proposed budget?

A 2,225   C 5,500

B 3,750   D 7,500

5 Which shows the sample space for this spinner?

A 1, 2, 3, 4, 5, 6, 7, 8

B 2, 4, 5, 6, 8, M, N, P

C 2, 3, 4, 5, 6, 7, 8, M, N, P

D 1, 2, 3, 4, 5, 6, 7, 8, M, N, O, P

6 Cory selected the lettered tiles shown below from a bag containing 120 tiles.

T	R	E	A	L
S	E	F	D	S

Based on the tiles he selected, he predicts that 35 of the total tiles are vowels. Is Cory's prediction reasonable?

A Yes, the total should be **about** 35.

B No, the total should be **about** 25.

C No, the total should be **about** 45.

D No, the total should be **about** 55.

Probability Review

Read each problem. Write your answers.

7 A box of 24 ice pops has 4 grape, 6 orange, 10 cherry, and the rest lemon.

   Part A: What is the probability of picking a grape pop from the box at random?

       Answer: _____

   Part B: What is the probability of **not** picking a cherry or a lemon pop from the box?

       Answer: _____

8 Look at this spinner.

   Part A: Make a list to show the sample space of possible outcomes when spinning the arrow on this spinner.

       Answer: _____

   Part B: What is the probability of getting an even number when spinning the arrow on the spinner?

       Answer: _____

UNIT 8 Probability

Read the problem. Write your answers.

9 Trevor inspected a box of 50 picture frames and found that 2 were damaged.

Part A: What is the probability that a picture frame selected at random will be damaged?

Answer: _____

Part B: The entire shipment contains 575 pictures frames. Trevor predicts that of these frames, a total of 50 will be damaged. Is Trevor's prediction reasonable?

Answer: _____

Explain how you know.

_____
_____
_____
_____

Empire State Mathematics
Supplemental Lessons

Writing Two-Step Equations

Indicator 6.A.3

equation   expression   equal sign (=)   variable

An **equation** is a number sentence that shows two **expressions** are equal. The expressions can be numeric expressions, algebraic expressions, or both. The **equal sign** shows the expressions are balanced.

$$5 + 7 = 2 + 10 \qquad x - 1 = 2x \qquad 3x = 15$$

> An algebraic expression contains a **variable** or a combination of numbers and variables.

Equations can use one operation to represent a relationship.

A roller coaster has c carts. Each cart can seat 6 people. A total of 72 people can ride the roller coaster at one time. Write an equation that can be used to find the number of carts on this roller coaster.

1. Identify the variable: c.

2. Write an expression using this variable: $6c$.

3. Set this expression equal to another expression to make a balanced equation: $6c = 72$.

> Expressions can show any operation or a combination of operations.
> Addition: $n + 8$
> Subtraction: $n - 5$
> Multiplication: $2n$, $2(n)$, $2 \times n$, $2 \cdot n$
> Division: $n \div 10$, $\frac{n}{10}$

Equations can also use more than one operation to represent a relationship.

Ramon read the same number of pages of a book each of the 5 weekdays last week. On that weekend, he read a total of 40 pages. Altogether, Ramon read 100 pages of the book last week. Write an equation that represents this situation.

1. Identify the variable. Let the variable p represent the number of pages Ramon read on each weekday.

2. Write an expression using this variable: $5p + 40$.

3. Set this expression equal to another expression to make a balanced equation: $5p + 40 = 100$.

> Key words can help to translate words into expressions.
> *Is* typically means equals.
> *More than* typically means add.
> *Less than* typically means subtract.
> *Product* means multiply.
> *Quotient* means divide.

Algebra

GUIDED PRACTICE

Try this sample multiple-choice problem.

S The deep end of a swimming pool is 2 feet less than 3 times the depth of the shallow end. The deep end is 7 feet. Which equation can be used to find the depth of the shallow end?

A $3 + s + 2 = 7$

B $3 \times s - 2 = 7$

C $2 - 3 + s = 7$

D $2 - 3 \times s = 7$

> This problem asks you to write an equation to find the depth of the shallow end of a swimming pool. The variable s represents this depth. The deep end is 2 feet less than 3 times the depth of the shallow end, or $3 \times s - 2$. The depth of the deep end is equal to 7. The correct answer is B.

INDEPENDENT PRACTICE

Read each problem. Circle the letter of the best answer.

1 Seven more than the product of k and 5 is -8. Which equation shows this?

A $5k + 7 = -8$ C $7k + 5 = -8$

B $5k \div 7 = -8$ D $7k \div 5 = -8$

2 A 3-pack of juice boxes costs $1.25. Ben bought a total of 30 juice boxes. Which equation can be used to find Ben's total cost, c, for these juice boxes?

A $c = 3 + 1.25 \times 30$

B $c = 3 \times 1.25 + 30$

C $c = (30 \times 3) \div 1.25$

D $c = (30 \div 3) \times 1.25$

3 The quotient of d minus 16 and 3 is 4. Which equation shows this?

A $\frac{d}{16} - 3 = 4$ C $\frac{d-16}{3} = 4$

B $\frac{d}{16} - 4 = 3$ D $\frac{d-16}{4} = 3$

4 A plumber charges $40 an hour and parts to fix a leak. The plumber charged Mr. Watson $84 to fix his sink, including $24 for parts. Which equation can be used to find the number of hours the plumber worked?

A $24h + 40 = 84$

B $40h + 24 = 84$

C $h + (24 + 40) = 84$

D $h \times (24 + 40) = 84$

5 During the first part of a trip, a car traveled 45 miles. Then it traveled for h hours at 60 miles per hour. The car went a total of 150 miles. Which equation shows this relationship?

A $45 + 60h = 150$

B $45h + 60 = 150$

C $45 \times (60 + h) = 150$

D $(45 + h) \times 60 = 150$

Algebra 5

GUIDED PRACTICE

Try this sample constructed response problem.

S Two classrooms have a combined area of 1,200 square feet. One classroom is 200 square feet larger than the other.

Part A: Write expressions for the areas of each classroom.

Smaller classroom: ____n____

Larger classroom: ___$n + 200$___

Part B: Write an equation that represents the total areas of both classrooms.

Answer: ___$2n + 200 = 1{,}200$___

> Part A asks you to write expressions for the areas of each classroom. Let the variable n represent the area of the smaller classroom. Since the larger classroom is 200 square feet greater, its area is $n + 200$. The correct answers are n and $n + 200$. Part B asks you to write an equation to represent the total areas. To do this, add the expressions from Part A. These will equal the total area of 1,200 square feet. The correct answer is $n + n + 200 = 1{,}200$ or $2n + 200 = 1{,}200$.

INDEPENDENT PRACTICE

Read the problem. Write your answers.

6 Denise's dog weighs 15 pounds more than 5 times the weight of her cat. Altogether, her dog and cat weigh 87 pounds.

Part A: If the cat weighs p pounds, write an expression for the weight of the dog.

Answer: _____

Part B: Write an equation that represents the total weight of the dog and cat.

Answer: _____

Algebra

INDEPENDENT PRACTICE

Read the problem. Write your answers.

7 A theater has two levels of seating. Level 1 has 80 more seats than level 2. Altogether, there are 320 seats on both levels.

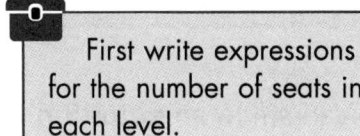
First write expressions for the number of seats in each level.

Part A: Let *s* represent the number of seats in level 2. Write an equation that can be used to find the number of seats in level 2.

Answer: _____

Part B: Next summer, the theater will add a third level. Level 3 will have twice as many seats as level 2. The total number of seats in the theater will increase to 560. Write a new equation to show this situation.

Answer: _____

Explain why your answer is correct.

_____
_____
_____
_____

Algebra

Solving Two-Step Equations

Indicator 6.A.4

solve   equation   variable   inverse operation   steps   isolate

You can **solve** an **equation** for the value of a **variable**. The value you find for the variable will make that equation true.

What is the value of n in the equation $n + 7 = 16$?

1. Identify the **inverse operation.** Subtraction is the inverse of addition.

2. Perform the inverse operation on both sides of the equation. Subtract 7 from each side of the equation:

$$n + 7 - 7 = 16 - 7$$
$$n = 9$$

The value of n is 9.

> Inverse operations "undo" each other. They are opposites.
>
> Addition and subtraction are inverse operations.
>
> Multiplication and division are inverse operations.

Some equations have two **steps,** or operations.

What is the value of d in the equation $3d - 5 = 13$?

1. Undo addition or subtraction. Add 5 to each side to undo subtraction:

$$3d - 5 + 5 = 13 + 5$$
$$3d = 18$$

2. Undo multiplication or division. Divide each side by 3 to undo multiplication.

$$\frac{3d}{3} = \frac{18}{3}$$
$$d = 6$$

The value of d is 6.

> To solve an equation, the same operation is **always** performed on both sides of the equation. This keeps the equation balanced as you **isolate,** or solve for the variable.

> Always undo the operation that is furthest from the variable first.

Algebra

GUIDED PRACTICE

Try this sample multiple-choice problem.

S What is the solution to the equation below?

$$\frac{t}{2} - 1 = 33$$

A 16

B 17

C 64

D 68

> This problem asks you to find the value of *t* that makes the equation true. Use inverse operations. First add 1 to each side to undo the subtraction: $\frac{t}{2} - 1 + 1 = 33 + 1$; $\frac{t}{2} = 34$. Next multiply each side of the equation by 2 to undo the division: $\frac{t}{2} \cdot 2 = 34 \cdot 2$; $t = 68$. The correct answer is D.

INDEPENDENT PRACTICE

Read each problem. Circle the letter of the best answer.

1 What is the first step in solving the equation below?

$$5k + 25 = 40$$

A add 25 to each side

B divide each side by 5

C multiply each side by 5

D subtract 25 from each side

2 What value of *s* makes $9 + 3s = 33$ true?

A 2 C 14

B 8 D 72

3 Bella bought a tennis racket and *c* cans of tennis balls for $75. The equation $2c + 65 = 75$ models this situation. How many cans of tennis balls did Bella buy?

A 5 C 70

B 20 D 85

4 What are the correct steps to use on each side of this equation when solving for *w*?

$$\frac{w}{6} - 3 = 36$$

A divide by 6, then add 3

B add 3, then multiply by 6

C subtract 3, then divide by 6

D multiply by 6, then subtract 3

5 Last week, Grayson read half the pages in a book. This week he read 22 more pages. Altogether, he read 96 pages. The equation below can be used to find *p*, the total number of pages in Grayson's book.

$$\frac{p}{2} + 22 = 96$$

What is the value of *p*?

A 70 C 148

B 140 D 236

Algebra

GUIDED PRACTICE

Try this sample constructed response problem.

S Derek answered 14 questions correctly on a quiz and earned a 7-point bonus. His score on the quiz was 91. The equation below can be used to find q, the point value of each correct question.

$$14q + 7 = 91$$

What is the value of q?

Show your work.

$$14q + 7 - 7 = 91 - 7$$
$$\frac{14q}{14} = \frac{84}{14}$$
$$q = 6$$

> This question asks you to find the value of q that makes the equation true. To do this, use inverse operations. Subtracting 7 is the opposite of adding 7, so subtract 7 from both sides. Then divide each side by 14 since division is the inverse operation of multiplication. The correct answer is 6.

Answer: ____6____

INDEPENDENT PRACTICE

Read the problem. Write your answers.

6 Look at this equation.

$$21 + \frac{y}{6} = 111$$

What value of y makes this equation true?

Show your work.

Answer: _____

10 Algebra

INDEPENDENT PRACTICE

Read the problem. Write your answers.

7 Look at this equation.

$$\frac{x + 3}{4} = 36$$

Part A: What are the correct steps to use when solving this equation for *x*?

Answer: _____

_____

> Use the inverse operations of division and addition.

Part B: What is the value of *x* that makes this equation true?

Show your work.

Answer: _____

Algebra 11

Solving Proportions

Indicator 6.A.5

proportion ratio cross multiply means extremes

A **proportion** is an equation that shows two ratios are equal.

$$\frac{2}{3} = \frac{8}{12}$$

This proportion says that the ratio 2 to 3 is the same as the ratio 8 to 12.

> A **ratio** is a comparison of two numbers.
>
> 3 teachers to 18 students
>
> $\frac{3}{18}$

Some proportions have variables that stand for missing numbers. You can solve a proportion to find the value of the missing number.

Three boxes of cereal cost $5. Ms. Quinn buys 12 boxes of this cereal. This proportion can be used to find y, the cost of 12 boxes of the cereal.

$$\frac{3}{5} = \frac{12}{y}$$

What is the value of y?

1. **Cross multiply.** $\qquad\qquad 3 \cdot y = 5 \cdot 12$
$\qquad\qquad\qquad\qquad\qquad\qquad 3y = 60$

2. Divide each side by 3 to find y. $\quad 3y \div 3 = 60 \div 3$
$\qquad\qquad\qquad\qquad\qquad\qquad\qquad y = 20$

The 12 boxes of cereal cost $20.

> The product of the **means** and the product of the **extremes** of a proportion are equal to each other.
>
>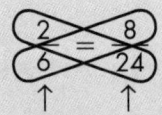
>
> ↑   ↑
> Means Extremes
>
> $2 \times 24 = 6 \times 8$
>
> $48 = 48$

> Cross multiplying is the same as setting the product of the means equal to the product of the extremes of a proportion.

12 Algebra

© The Continental Press, Inc. Do not duplicate.

GUIDED PRACTICE

Try this sample multiple-choice problem.

S On a table, 30 mosaic tiles are used for every 4 square feet of space. Jesse wants to decorate a 20-square foot table with mosaic tiles. This proportion can be used to find the number of mosaic tiles Jesse will need.

$$\frac{30}{4} = \frac{m}{20}$$

What is the value of *m*?

- **A** 46
- **B** 60
- **C** 120
- **D** 150

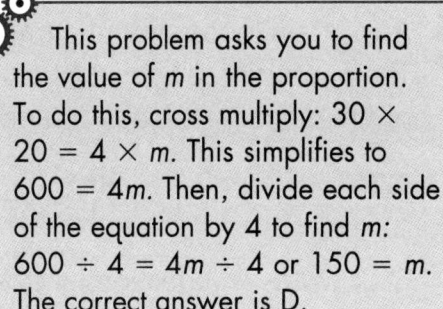

This problem asks you to find the value of *m* in the proportion. To do this, cross multiply: $30 \times 20 = 4 \times m$. This simplifies to $600 = 4m$. Then, divide each side of the equation by 4 to find *m*: $600 \div 4 = 4m \div 4$ or $150 = m$. The correct answer is D.

INDEPENDENT PRACTICE

Read each problem. Circle the letter of the best answer.

1 The ratio of boys to girls at a camp is 4 to 5. The proportion below can be used to find the number of girls at the camp with 20 boys.

$$\frac{4}{5} = \frac{20}{g}$$

How many girls are at the camp?

- **A** 20
- **B** 21
- **C** 24
- **D** 25

2 To make a smoothie, Lucy uses 1 part crushed ice to 2 parts yogurt. Lucy wrote the proportion $\frac{1}{2} = \frac{c}{3}$ to find how many cups of crushed ice she will need to make a smoothie using 3 cups of yogurt. How many cups of crushed ice will Lucy need?

- **A** $\frac{2}{3}$
- **B** $1\frac{1}{2}$
- **C** 3
- **D** 6

3 The ratio of a rectangle's length to its width is 2 to 3. The proportion below can be used to find the width of a rectangle with a length of 36 inches.

$$\frac{2}{3} = \frac{36}{w}$$

What is the width of this rectangle?

- **A** 24 inches
- **B** 37 inches
- **C** 54 inches
- **D** 60 inches

4 While exercising, Andy counts 25 heartbeats in 10 seconds. The proportion $\frac{25}{10} = \frac{h}{60}$ can be used to find his number of heartbeats in 60 seconds. What is the value of *h*?

- **A** 75
- **B** 150
- **C** 750
- **D** 1,500

Algebra

GUIDED PRACTICE

Try this sample constructed response problem.

S A toy train loops around a track 10 times in 4 minutes. The proportion below can be used to find the number of times the train loops around the track in 10 minutes.

$$\frac{10}{4} = \frac{t}{10}$$

Zack writes an equation by cross multiplying to find the value of t.

Part A: What equation results from cross multiplying?

   Answer: __100 = 4t__

Part B: What is the value of t in this proportion?

   Answer: __25__

> Part A asks you to write the equation that results from cross multiplying in the proportion. To cross multiply, set the product of the means equal to the product of the extremes: $10 \times 10 = 4 \times t$. Then simplify. The correct answer is $100 = 4t$. Part B asks you to find the value of t in the proportion. To do this, use the equation you wrote in Part A. Divide each side by 4: $100 \div 4 = 4t \div 4$. The correct answer is 25.

INDEPENDENT PRACTICE

Read the problem. Write your answers.

5 In a survey, 30 out of 50 students said they would vote to have a school dance in the gymnasium. The proportion below can be used to find the number of students in the entire school who would be expected to vote to have the school dance in the gymnasium.

$$\frac{30}{50} = \frac{v}{600}$$

What is the value of v?

Show your work.

Answer: _____

14   Algebra

INDEPENDENT PRACTICE

Read the problem. Write your answers.

6 A caterer charges $30 for a pan of lasagna that feeds 6 people. The caterer uses the proportion below to determine the amount to charge for a pan of lasagna that feeds 15 people.

$$\frac{30}{6} = \frac{a}{15}$$

Part A: What two parts are the means and what two parts are the extremes in this proportion?

Means: _____

Extremes: _____

Part B: What amount does the caterer charge for the pan of lasagna that feeds 15 people?

Show your work.

> Cross multiply to make a new equation. Then solve for the variable.

Answer: _____

Lesson 4: More Coordinate Geometry

Indicators 6.G.10, 11

coordinate plane   ordered pair   coordinates   x-value   y-value
x-axis   y-axis   origin   quadrants

Points are plotted on a **coordinate plane.** The location of each point is named by an **ordered pair** of numbers called **coordinates.**

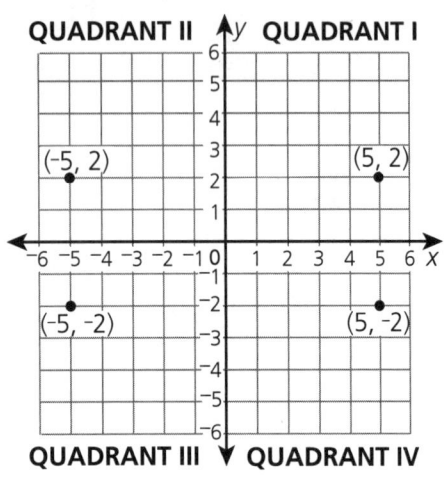

The first coordinate, or **x-value,** names the distance a point is left or right from the origin. The second coordinate, or **y-value,** names the distance the point is up or down from the origin.

> A coordinate plane has two axes. The **x-axis** is horizontal. The **y-axis** is vertical. The point where both axes intersect is the **origin.** The coordinates of the origin are (0, 0).

> There are four **quadrants** on a coordinate plane.

You can use a coordinate plane to find the area of a rectangle.

What is the area of rectangle HJKL?

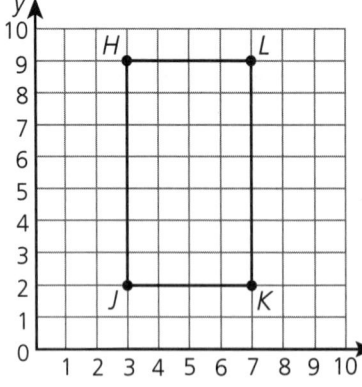

1. Find the length of the rectangle: J = (3, 2) and K = (7, 2). Subtract the x-values: 7 − 3 = 4.

2. Find the width of the rectangle: H = (3, 9) and J = (3, 2). Subtract the y-values: 9 − 2 = 7.

3. Multiply the length and width: 7 × 4 = 28 square units.

> Sometimes a coordinate plane shows only one of the four quadrants.

> Length is the difference between the x-coordinates.
> Width is the difference between the y-coordinates.

Geometry

GUIDED PRACTICE

Try this sample multiple-choice problem.

S What is the area of this composite figure?

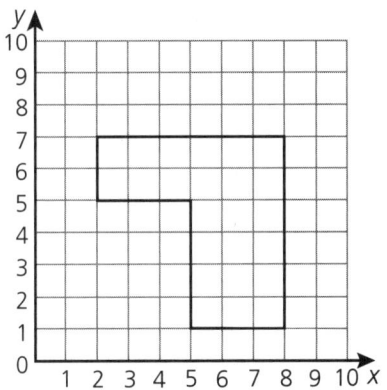

- **A** 18 sq units
- **B** 24 sq units
- **C** 30 sq units
- **D** 36 sq units

> This problem asks you to find the area of a composite figure. Divide the figure into two rectangles:
>
> Find the area of each. Rectangle 1 is 3 units long and 2 units wide. The area of rectangle 1 is 3 × 2 = 6 sq units. Rectangle 2 is 3 units long and 6 units wide. The area of rectangle 2 is 3 × 6 = 18 sq units. Add the areas of rectangles 1 and 2: 6 + 18 = 24 sq units. The correct answer is B.

INDEPENDENT PRACTICE

Read each problem. Circle the letter of the best answer.

Use this coordinate plane to answer questions 1–4.

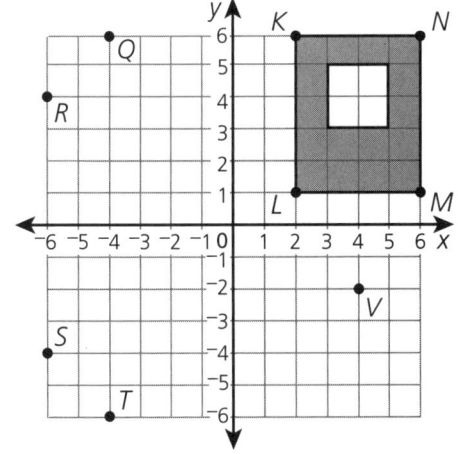

1 What are the coordinates of point *V*?

- **A** (-4, 2)
- **B** (-2, 4)
- **C** (2, -4)
- **D** (4, -2)

2 Which point has the coordinates (-6, -4)?

- **A** point *Q*
- **B** point *R*
- **C** point *S*
- **D** point *T*

3 What is the area of rectangle *KLMN*?

- **A** 16 sq units
- **B** 20 sq units
- **C** 24 sq units
- **D** 36 sq units

4 What is the area of the shaded region of rectangle *KLMN*?

- **A** 4 sq units
- **B** 16 sq units
- **C** 18 sq units
- **D** 20 sq units

Geometry

GUIDED PRACTICE

Try this sample constructed response problem.

S Polly plotted rectangle ABCD on this coordinate plane.

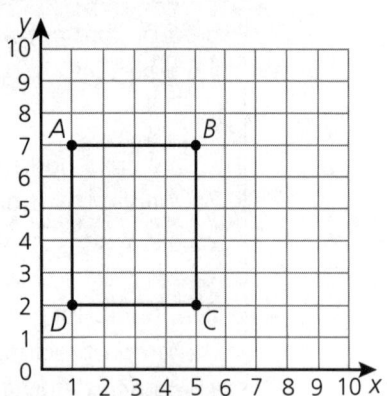

 Part A asks you to find the length of side BC. First identify the coordinates of points B and C: B = (5, 7) and C = (5, 2). Subtract the y-values to find the length: 7 − 2 = 5. The correct answer is 5 units. Part B asks you to find the area of the rectangle. From part A, you know the width is 5 units. Find the length: D = (1, 2) and C = (5, 2). Subtract the x-values: 5 − 1 = 4. The length is 4. Multiply to find the area: 5 × 4 = 20. The correct answer is 20 sq units.

Part A: What is the length of the side labeled BC?

Answer: __5 units__

Part B: What is the area of this rectangle?

Answer: __20 sq units__

INDEPENDENT PRACTICE

Read the problem. Write your answers.

5 Carlos will plot the point Z = (3, -2) on this coordinate plane.

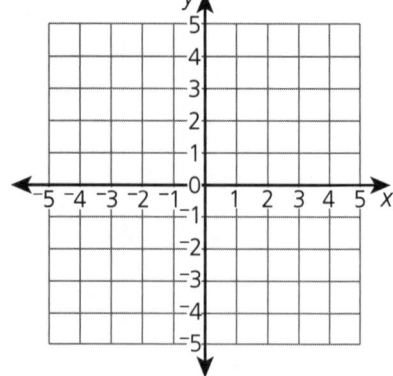

Part A: What quadrant is point Z in?

Answer: _____

Part B: On the coordinate plane at the right, plot and label point Z.

18 Geometry

INDEPENDENT PRACTICE

Read the problem. Write your answers.

6 Look at the figure on this coordinate plane.

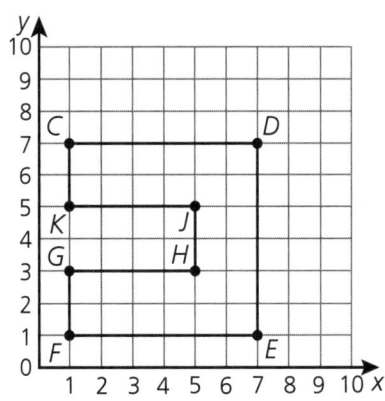

Part A: Which point on this figure has the coordinates (1, 7)?

Answer: _____

Part B: What is the area of this figure?

Divide the figure into rectangles.

Answer: _____

Explain how you found your answer.

_____

_____

_____

_____

Geometry 19

LESSON 5

Compound Events

Indicators 6.S.9, 11

compound event   organized list   sample space   event
probability   fundamental counting principle

A **compound event** combines two or more single events. An **organized list** or **sample space** can show all the possible outcomes of a compound event.

Veronica spins the arrow on this spinner and flips a coin. What are all the possible outcomes of this compound event?

1. Identify the possible outcomes for the spinner and the possible outcomes for the coin. The spinner can land on 1, 2, or 3. The coin can land on heads or tails.

2. Combine these outcomes to make an organized list.

Spinner	Coin
1	heads
2	heads
3	heads
1	tails
2	tails
3	tails

> An **event** is something that can happen.
>
> A **sample space** is the set of all possible outcomes.

> To make an organized list, combine all the outcomes from one set with all the outcomes from the other set, in order.

To find the **probability** of a compound event, multiply the probability of one event by the probability of the other event.

$$P(A \text{ and } B) = P(A) \times P(B)$$

The **fundamental counting principle** is a multiplication rule that can be used to find the number of possible outcomes for a compound event.

How many total possible outcomes are there when flipping a coin and rolling a six-sided number cube with the numbers 1 through 6?

1. Identify the number of possible outcomes for each event: flipping coin, 2; rolling number cube, 6.

2. Multiply: 2 × 6 = 12 possible outcomes.

> The probability (P) of an event is the chance it will happen. Probabilities are numbers between 0 and 1 expressed as a fraction, decimal, or percent.
>
> $P = \dfrac{\text{Favorable outcomes}}{\text{Possible outcomes}}$

Probability

GUIDED PRACTICE

Try this sample multiple-choice problem.

S Carter flips a coin three times. What is the probability that the coin lands on heads all three times?

A $\frac{1}{8}$ C $\frac{1}{3}$

B $\frac{1}{6}$ D $\frac{1}{2}$

> This problem asks you to find the probability of a compound event with three separate events. The probability of a coin landing on heads once is $\frac{1}{2}$. Multiply to find the probability of the three events: $\frac{1}{2} \times \frac{1}{2} \times \frac{1}{2} = \frac{1}{8}$. The correct answer is A.

INDEPENDENT PRACTICE

Read each problem. Circle the letter of the best answer.

1 Kerry will ride the bus two times this week. Each day, the bus will either be early (E), on time (O), or late (L). Which list shows the sample space of all possible outcomes for this event?

A E, O, L

B EE, OO, LL

C EO, EL, OE, OL, LE, LO

D EE, EO, EL, OE, OO, OL, LE, LO, LL

2 Linda spins the arrow on this spinner twice.

How many total possible outcomes are there for this compound event?

A 6   C 36

B 12   D 64

3 A restaurant offers these pastas: spaghetti (S), ziti (Z), linguini (L). These sauces are offered: tomato (T), meat (M). Which list shows all possible outcomes for picking one pasta and one sauce?

A S, Z, L, T, M

B ST, ZM, LT

C ST, SM, ZT, ZM, LT, LM

D SZ, SL, ST, SM, ZL, ZT, ZM, LT, LM

4 Craig rolls a six-sided cube with the numbers 1 through 6 a total of three times. What is the probability that his first roll is an even number, his second roll is an odd number, and his third roll is the number 4?

A $\frac{1}{24}$ C $\frac{1}{8}$

B $\frac{1}{18}$ D $\frac{1}{6}$

Probability

GUIDED PRACTICE

Try this sample constructed response problem.

S Mandy flips a quarter four times.

Part A: How many total possible outcomes are in the sample space for this compound event?

   Answer: _____16_____

Part B: What is the probability that all four flips will result in the same outcome?

   Answer: _____$\frac{1}{8}$_____

> Part A asks you to find a total number of possible outcomes. For each coin flip, there are two possible outcomes: heads or tails. Use the fundamental counting principle: $2 \times 2 \times 2 \times 2 = 16$. The correct answer is 16. Part B asks you to find the probability that all four flips will have the same result, either all heads or all tails. First, find the probability that all are heads: $\frac{1}{2} \times \frac{1}{2} \times \frac{1}{2} \times \frac{1}{2} = \frac{1}{16}$. The probability that all are tails is also $\frac{1}{16}$. The probability that all four flips result in all heads or all tails is $\frac{1}{16} + \frac{1}{16} = \frac{1}{8}$. The correct answer is $\frac{1}{8}$.

INDEPENDENT PRACTICE

Read the problem. Write your answers.

5 In basketball, Will can either get the ball in the basket (B) or miss the basket (M) on each attempt. Will has three attempts.

Part A: List the possible outcomes for this compound event.

   Answer: _____

Part B: What is the probability that Will gets the ball in the basket in all three attempts?

   Answer: _____

22 Probability

INDEPENDENT PRACTICE

Read the problem. Write your answers.

6 Jamar flips a coin and rolls a six-sided cube with the numbers 1 through 6.

 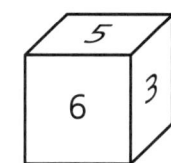

Part A: How many outcomes are in the sample space for this compound event?

Answer: _____

List all possible outcomes in the sample space.

Answer: _____

Part B: What is the probability that the coin lands on tails and Jamar rolls a number less than 3?

> Find the probability of rolling a number less than 3. Find the probability of the coin landing on tails. Multiply.

Answer: _____

Probability

Dependent Events

Indicator 6.S.10

dependent   independent

Two events are **dependent** if the outcome of the first event affects the outcome of the second event.

Two events are **independent** if they have no effect on one another.

A box contains these pens.

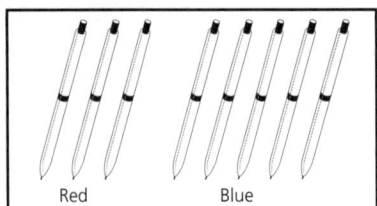

Shawn will pick 2 pens from the box. What is the probability that both pens he picks will be blue?

1. Find the probability the first pen will be blue.

    $\frac{\text{Number of favorable outcomes}}{\text{Number of possible outcomes}} = \frac{5 \text{ blue pens}}{8 \text{ total pens}} = \frac{5}{8}$

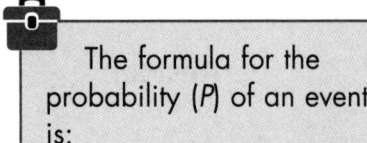

The formula for the probability (P) of an event is:

$P = \frac{\text{Favorable outcomes}}{\text{Possible outcomes}}$

After Shawn picks the first pen, only 7 pens remain. So the number of possible outcomes for the second event is smaller. This shows that the second event is dependent on the first event.

2. Find the probability the second pen will be blue.

    $\frac{\text{Number of favorable outcomes}}{\text{Number of possible outcomes after first event}} = \frac{4 \text{ blue pens left}}{7 \text{ total pens left}} = \frac{4}{7}$

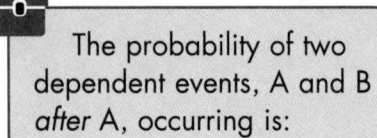

The probability of two dependent events, A and B after A, occurring is:

$P(A) \times P(B)$

3. Multiply the probability of the first event by the probability of the second event: $\frac{5}{8} \times \frac{4}{7} = \frac{20}{56} = \frac{5}{14}$.

The probability that 2 blue pens will be picked from the box is $\frac{5}{14}$.

GUIDED PRACTICE

Try this sample multiple-choice problem.

S A cooler contains 6 cans of iced tea and 3 cans of juice. Maggie picks one can of juice from the cooler. Later, she picks another can from the cooler. What is the probability that the second can will be iced tea?

A $\frac{1}{2}$ C $\frac{2}{3}$

B $\frac{5}{8}$ D $\frac{3}{4}$

> This problem asks you to find the probability of a second event occurring after the first event has happened. After Maggie picks 1 juice, there are 2 juices and 6 iced teas left. This totals to 8 cans in all. The probability that she picks an iced tea from the 8 remaining cans is $\frac{6}{8}$ or $\frac{3}{4}$. The correct answer is D.

INDEPENDENT PRACTICE

Read each problem. Circle the letter of the best answer.

1 There are 10 containers of yogurt in Javier's refrigerator. Half are peach and half are strawberry. Javier picks 2 yogurts without looking. The first is strawberry. What is the probability that the second flavor he picks is also strawberry?

A $\frac{2}{5}$ C $\frac{1}{2}$

B $\frac{4}{9}$ D $\frac{5}{9}$

2 A game contains the shapes shown below.

△ △ △ △ ◇ ◇ ◇ □ □

Luke picks two shapes without looking. The first shape he picks is a ◇. What is the probability that the second shape he picks is a △?

A $\frac{2}{9}$ C $\frac{4}{9}$

B $\frac{1}{3}$ D $\frac{1}{2}$

3 Candice has 4 gold bracelets and 2 silver bracelets. She picks a gold bracelet to wear. Then she picks another bracelet without looking. What is the probability that the second bracelet she picks is silver?

A $\frac{1}{3}$ C $\frac{1}{2}$

B $\frac{2}{5}$ D $\frac{2}{3}$

4 A box of instant oatmeal contains these flavors: 3 maple, 2 cinnamon, and 3 raisin. Mayuko picks a flavor of oatmeal without looking. Then her sister picks one. What is the probability that **both** flavors are raisin?

A $\frac{3}{28}$ C $\frac{9}{64}$

B $\frac{1}{9}$ D $\frac{3}{8}$

GUIDED PRACTICE

Try this sample constructed response problem.

S A pile contains these cards.

| 3 | 2 | 6 | 5 | 9 | 8 |

Ethan picks the card with the number 5. Then he picks two more cards without looking.

Part A: What is the probability that the second card Ethan picks is a 3?

Answer: $\frac{1}{5}$

Part B: What is the probability that Ethan picks a card with a 5, then a card with a 3, and then an even-numbered card?

Answer: $\frac{1}{40}$

> Part A asks you to find the probability of picking a 3 after a 5 has been picked. After Ethan picks the first card, only 5 cards are left. One of these cards is 3, so the correct answer is $\frac{1}{5}$. Part B asks you to find the probability of three dependent events. Find the probability of each event and multiply them. The probability of the first event is $\frac{1}{6}$ and of the second event, $\frac{1}{5}$. For the third event, only 4 cards are left. Of these, 3 are even. So the probability of the third event is $\frac{3}{4}$. The probability of all events is $\frac{1}{6} \times \frac{1}{5} \times \frac{3}{4} = \frac{3}{120} = \frac{1}{40}$. The correct answer is $\frac{1}{40}$.

INDEPENDENT PRACTICE

Read the problem. Write your answers.

5 Vern has 16 toy cars. Of these, 3 are blue, 8 are red, and 5 are yellow. Vern picks two toy cars to give to a friend. The first toy car he picks is red.

What is the probability that the second toy car Vern picks is blue?

Answer: _____

Explain why your answer is correct.

_____
_____
_____
_____

26 Probability

© The Continental Press, Inc. Do not duplicate.

INDEPENDENT PRACTICE

Read the problem. Write your answers.

6 A bowl of mints contains 9 peppermints and 12 spearmints. Dwayne will pick two mints from the bowl.

Part A: What is the probability that **both** mints will be peppermints?

Show your work.

Answer: _____

Part B: Dwayne decides to pick a third mint from the bowl. What is the probability that the third mint will be spearmint?

Find the total number of mints left in the bowl after the first two picks.

Answer: _____